AF602150

Fishery Resources of Goa and Marketing

A Geographical Approach

The Authors

Dr. Dadapir Moulasab Jakati completed his Post-Graduation in Geography from Karnatak University Dharwad and M.Sc. in Ecology and Environment from SMU. He obtained his M. Phil in Resource Geography and Doctoral Degree in Resource and Marketing Geography.(May 2012) Presently, he is a Associate Professor of Geography and has been teaching in the Under-Graduate Department of Geography, St. Xavier's College Mapusa. Bardez, Goa since 1995. He is engaged in research in the specializations of Resource Geography, and Marketing Geography. His research papers are published in the known professional journals. He is also presented the research papers in the both National and International Conferences in India. Beside these, he is He is Promoter of Environmental Education of IIEE. New Delhi. He is represented as member of B.O.S. in Geography and Environmental Sciences at Goa University, Goa Life member of Geographers Association Goa. and Ph.D. Guide/ Research Supervisor of JJT. Universty Rajasthan. The author's trust is to deeply involve/ contribute to take up the research in different fields of Geography. He has been elected as Member to the Court of Goa University, Goa. from constituency no.4.

Dr. Aravind A. Mulimani (b.1964) has to his credit Post-Graduation in Geography (1989), with First Rank and GOLD MEDAL from Karnatak University Dharwad. He obtained his M. Phil in Resource Geography and Doctoral Degree in Marketing Geography. Presently, he is a Professor of Geography and has been teaching in the Post-Graduate Department of Geography, Karnatak University, Dharwad since 1990. He is engaged in research in the specializations of Resource Geography, Settlement and Marketing Geography. Dr. Mulimani has guided for M. Phil and Ph.D's. His research papers are published in the known professional journals. He is also presented the research papers in the both National and International Conferences not only in India but also in abroad. Beside these, he is also delivered Keynote address and Invited Lectures in many National Seminars /conferences in different parts in India. He is acting as Vice-President of Deccan Geographical Society of India, Pune since, 2004. The author's trust is deeply involve/ contribute to take up the research in different fields of Geography.

Fishery Resources of Goa and Marketing

A Geographical Approach

Dadapir Moulasab Jakati
Aravind A. Mulimani

2020
Scholars World
A Division of
Astral International Pvt. Ltd.
New Delhi – 110 002

ISBN 9789390371297 (Int. Edition)

Scholars World
A Division of
Astral International Pvt. Ltd.
– ISO 9001:2015 Certified Company –
4736/23, Ansari Road, Darya Ganj
New Delhi-110 002
Ph. 011-4354 9197, 2327 8134
E-mail: info@astralint.com
Website: www.astralint.com

Digitally Printed at : **Replika Press Pvt. Ltd.**

Dedication

"Affectionately Dedicated to My Beloved Parents."
My Geography Teacher Dr. C.S. Unakall,
Principal Late.Rev.Canon. Antimo Gomes.

Acknowledgement

At the outset I sincerely express my deep sense of gratitude to the almighty who has given me strength, patience and enlightenment in completion of the research work.

It is my proud privilege and great honor to express my deep sense of gratitude to my Research Guide and Supervisor Dr. Aravind A. Mulimani, Professor and Chairman, Department of Studies in Geography, Karnatak University, Dharwad, for his able guidance, unstinted encouragement, inspiring discussion and great support throughout this work.

Indeed I am grateful to the President and Secretary of the Diocesan Society of Education Management, Panaji, Goa. Rev Fr. Dr Jeronimo Desilva Acting Principal and Rev Fr. Dr. Walter D'Sa, Principal of St. Xavier's college for their encouragement and cooperation throughout my research.

My deep sense of gratitude to Shri Bhaskar G. Nayak Ex Director of Higher Education, Government of Goa and Ex -President D.S.E. Management, Rev. Canon Antimo Gomes, for their blessings, encouragement and great support in my academic career.

I must also express my sincere thanks to all the officials of Directorate of Fisheries, Government of Goa; Directorate of Census Operations; Directorate of Planning and Statistics Evaluation; Department of Industries and Mines; Department of Tourism; Goa Mineral Ore Exporters Association; Fishery Survey of India, Mormugao Indian Council of Agricultural Research Complex, Old Goa; Marine Products Exporters Development Association (MPEDA); Exporters Inspection Agency CIA; Karnatak University Library Staff; Goa University; National Institute of Oceanography, Donapaula, Goa; St. Xaviers College Library.

I personally extend my sincere thanks to Dr. Subramanian, Dr. Mohanta, Scientists, Fishery Section, Indian Council of Agricultural Research, Old Goa, for their valuable suggestions and Shri Raghuram, Librarian, ICAR, for necessary help and cooperation.

My special thanks are due to my all the colleagues and friends of St. Xavier's college, Mapusa, Goa for their best wishes and blessings.

I shall be failing in my duty if I don't express heartfelt gratitude to my father Moulasab Jakati, mother Late Khairunisa Jakati, late Grandparents, uncles for their blessings and support.

I am ever grateful to my beloved wife Farzana who stood by me in all times and helped me a lot in completion of my research work. I also acknowledge my lovely daughters Ms. Ruhinaaz and Ms. Rida for their loving support.

I sincerely thank Dr. Sawant, Principal of S.P. Chaugule College, Dr. R.B. Patil, M.E.S. College, Dr. Oscar D Mello for their valuable co-operation and help. I am very thankful to Ms. Sumata and Mr. Subodh, Post Graduate students, eoinformatic Centre, Goa, for impressive, cartographic work. Dr.Prabirkumar Rath and Dear geographers of Goa for their necessary support and guidance in my academic career.

Dr. Dadapir Moulasab Jakati
Dr. Aravind A. Mulimani

Foreword

I am happy to write the forward to the book titled. "***Fisheries Resources of Goa and Marketing – A Geographical approach***" written by Dr. Dadapir Moulasab Jakati.

Goa is blessed with rich coastal areas that enable distribution of fisheries at specific places. This would be an excellent piece of information rooted through geographical perspectives by unfurling the fishery resources in Goa.Dr. Jakati has undoubtedly assimilated the significance of fish trade and it resources in his study. According to him, there is a large scope for fishery in Goa, which is increasing rapidly. Our future prospects lies in the present, therefore this research would be important in solving various issues associated with fish landings and trade.

The author has examined the development of marketing in fishery resources in the state. This book deals with the study of landings, major as well as minor in different talukas and analyses the types and the factors that influence the resources and the diversity. It focuses on local and International trade differences. The author has also brought out the tribulations faced by the fisher community and has laid the observations and recommendations.

This book would help to configure the fishing resources, its distribution and the economic conditions in Goa to an extent. This book would serve as a piece of information with framework in minds for future development in fishing industry.

It is hope that this publication, "Fisheries Resources of Goa and Marketing – A Geographical approach" will evoke considerable interest in students and researchers of Geography and will also serve as stimulating reading for the general reader.

Pernem

Dr. Filipe Rodrigues e Melo
Principal
S.S.A.Govt. College of Arts & Commerce,
Pernem, Goa and Head, Research Centre in Commerce

Preface

Goa is one of the states as far as fishery resources are concerned. Fish assumes scientific significance to the people of Goa, as it forms one of the important items of daily food of more than 90 percent of total population. Fishing serves as a means of livelihood to a large number of people of Goa. About 60 percent of the total active fishermen are engaged in marine fishing, while the remaining 40 percent in inland fishing activities.

In addition to fishing, a number of ancillary and subsidiary activities, have grown around fish harvesting, and help in earning considerable amount of foreign exchange through exports of various fish and fish products and domestic income by supplying fish to neighbouring states in the country.

The pace of fishing industry in Goa was somewhat slow, because the state could not get benefits of the first and second five year plan as it was under the Portuguese rule. However, after the liberation progress has been quite encouraging and shows rich potential for development of the fishing industry in Goa. In recent years, formers in the coastal region are turning away from the traditional agriculture and taking up aquaculture and therefore, the industry is growing rapidly. The scope for development of fisheries in the state is unlimited because the field seems still a virgin one, which urged me to undertake study of such a type of resource. As a result, the topic "Development of Marketing of Fishery Resources of Goa" has been selected.

The present confined investigation deals with the geographical perspective of Development of Marketing of Fishery Resources of Goa, and has been arranged systematically into seven chapters in order to understand the entire work carried out. The chapterization is as follows.

An attempt has been made in the first chapter, to give a brief introduction, definitions, concept of resources with different views and its types. Importance in Indian and Global scenario, with a brief review of literature. The marketing and its significance is also dealt with in accordance with the objectives of the study. In addition, hypothesis, methodology and the study area along with research design have been discussed in this chapter.

The second chapter, related to the spatial distribution of fish landing centres, focuses on brief introduction to spatial distribution, coastal geomorphology, geological characteristics, i.e., coastal morphology, fluvial farms, fluvial marine farms, marine farms, acolian farms, type of fish landing centres have been analysed taking taluka as a unit of analysis. The nearest neighbouring techniques are employed to understand the spatial pattern. The temporal distribution of major and minor fish landing centres are also paid due attention.

The fishery resource potentialities of Goa are the base chapter and hence an attempt has been made in the third chapter to discuss different aspects related to the fisheries. The introduction, factors influencing development of fishery resources, kinds of fisheries, fish species diversity, methods of fishing related to both marine as well as inland. The talukwise distribution of fishing gears and temporal trend of fishing gears and crafts have been analysed. Inland fisheries and marine fisheries have been discussed in detail.

The researcher has chosen to study the marketing activities of fishery resources in the fourth chapter. The present study has paid special attention to study of international trade along with local trade. The local trade system in different markets are also highlighted.

The fifth chapter deals with planning and strategies with special reference to problems of fisherman community, traders, the role of Government of Goa, private agencies and other leading organizations in sustainable growth of fishery resources and development of fishing industry in Goa as well as trade in fisheries.

The last chapter is a concluding chapter confined to a brief summary of the entire study. The specific observations and recommendations for spatial development of fishery resources and its marketing has been discussed.

Dadapir Moulasab Jakati

Aravind A. Mulimani

Contents

Chapter-1
INTRODUCTION

1.1 SIGNIFICANCE OF GEOGRAPHY

Geography is a multidisciplinary subject, and is gaining much importance. It has attracted not only by the academicians and planners, but also administrators, which provides enough scope to take up research from the micro level to the macro level of any geographical regions in the world. As a result, such a kind of research output helped the planners as well as administrators to a great extent for the development of any region. The latest technology is also responsible for fast growing knowledge in the field of subject. Hence, the discipline plays a significant role in the modern world.

As mentioned by Husain (1990) Geography has a longer genealogy than any other science or discipline on the earth surface. It has witnessed fluctuations of various phases of its status in the world.

Geography has become an extremely varied, versatile and high empirical discipline, and its area of study has grown a great deal since its inception. Harvey (1969) opines that geography is concerned with the description and explanation of the areal differentiation on the earth's surface.

Within the field of Geography, the Economic Geography has witnessed remarkable changes from its inception, and has also paved the way for emergence of sub-branches, and has taken all of them into its fold. As a result, the resources, geography, marketing, commercial, transportation, rural settlements, urban settlements, regional geography and regional planning, etc. have emerged. These branches are responsible for vibrant economic activities as also for economic development of the region. It has reflected on socio-economic activities that have provided enough employment opportunities to the people of the region. The resources are the base of the economy, and play a vital role in determining the life of the people.

The materials which are provided by the nature as free gifts to the human society are rich endowments. The fertile land and soil, water, air, natural vegetation, solar energy and minerals are the endowments which existed on the earth's surface even before the emergence of man. Man is capable of exploiting the resources to cater to the needs of the society and satisfy the human wants. Therefore, the resources are significant not only today but also for future generations.

Oxford English Dictionary defines resource, as a means of supply of some wants or deficiency, a stock or reserves upon which one can draw whenever necessary.

Ginsburg (1957) has considered natural resources as "all the freely given material phenomena of nature within the zone of man's activities plus the additional non-material quality of situation or location." The scholar has considered the association of land, water, air and situation in a single area, as it is a resource base or "resource endowment". In the words of Zimmerman (1951), "much of the environment is in fact composed of natural shift awaiting awareness of its possibilities and the development of technologies which can exploit it effectively".

Harvey (1973) describes resources as social and technological appraisal related to objects. The society oriented and relative value of natural resource is particularly prominent in economic cum geographical considerations.

1.2 CONCEPT OF RESOURCES

The concept of resource is dynamic and hence it changes with advancement of science and technology and culture, because the resources are identified on the bases of their usefulness to man. **Zimmerman** (1951) explains that the environment and part of the environment are not considered as resources unless they are capable of satisfying the needs of mankind. According to him, the chief criterion of resources is their availability for human use, and not their mere physical presence. Availability in turn depends on human wants and abilities to utilize these resources.

The word resource is a subjective concept. Coal, simply because of its physical composition is not a resource but it becomes one as man possesses wants which can be satisfied by releasing its stored up energy and turning it into heat. Most of the materials, lying on the earths surface practically unused by man, became subjective. For example, fertile soil, water, air, animal, plants, etc. Thus to be considered as a resource, endowment must be brought into the relationship of man. The resource concept is, therefore, relative also.

Further, endowments must be made to function so that they can be employed to satisfy human wants. When endowments are turned into useful things to satisfy human wants, they assume the status of resources. It is rightly said, "resources are not, they are to be", it is also said, "resources is what resource does." Hence the resource concept is also functional. For example: fertile soil. Waterfall, sunshine, air minerals etc., are brought into the relationship of man for practical uses in order to satisfy growing human needs. Thus living and non living materials are becoming functional in their nature while satisfying never ending human wants.

In a broader sense, resources may be defined as the means of satisfying human wants. In fact, any substance that possesses the property of benefiting man or leading to human welfare is a resource. Soil is a resource as crops can be cultivated from it. Similarly air, water, minerals, etc., are resources as they perform the function of satisfying human wants. It is this function of a substance of satisfying human wants that makes it a resource.

Prof. Zimmerman (1951) defines, "the word resource does not refer to a thing or a substance but to a function which a thing or a substance may perform or to an operation in which it may take part, namely, the function or operation of attaining a giver end such as satisfying a human want. An important quality of each resource is that it can satisfy a number of wants. Any substance, which can give from utility through its products to satisfy human wants, is a resource. Thus, resources have alternative uses. The domesticated animals, like cattle and sheep provide not only meat, but milk wool, hides, skins and bones etc. Also, each of which has a variety of uses. Similarly, other materials as well, i.e., water, fisheries, natural vegetation, etc. Utility to man is the main characteristics of a resource. Strictly speaking, therefore a resource is a natural resource, source of flow of substances from which a number of materials can be obtained for satisfaction of human wants.

Mankind is wholly dependent upon the inorganic and organic spheres of the globe. He uses atmosphere, lithosphere and hydrosphere as resources which are mobilized for human needs. Man as a supreme commander of exploitation of his resources constantly changes his attitudes, techniques and decisions from time to time.

Growing human population is responsible for increasing demand for resources, which is coupled with the rising standard of living. Increased standard of living demands high purchasing power and consumption level. As a result, pressure on resources increases, for example: in U.S.A., average level of resources use is greatest in the world. The country which exerts pressure on its resources increases them several folds. World population growth does not run parallel to resource supply, and thus the gap between demand and supply is widening with the advancement of science and technology.

1.3 DISCOVERY AND DEVELOPMENT OF RESOURCES

It is said that before creating man, God created the resources. It was left to man to discover, develop and exploit these resources for his own use. Thus air, water, soil, animals, plants, etc., were all created much before the creation of man. During the period of chemical and biological evolution on the earth between 500 millions years to 2000 millions years, these resources provided the initial base on which primitive man worked directly to obtain goods for satisfying his basic needs. He collected fruits from forests, caught fish from water, killed animals for their meat. With his intelligence, ingenuity, knowledge and experience, he learnt to cultivate the soil to grow food. He also made simple tools and implements to carry on these activities more effectively.

The Stone Age man did not know value of metals. But these were the very substances from which he made weapons and implements and built vast empires in the bronze and copper age. The Bedouins (Bedouins) of West Asia used to ride over the desert waste lands without the slightest idea of the vast wealth of soil that was lying concealed under their feet. Today there is a great power struggle between the super powers of the world to gain access to the same waste lands. The sun has been providing unlimited amount of energy through the ages, and nobody bothered about it. Today continuous efforts are being made all over the world to harness the solar energy.

1.4 TYPES OF RESOURCES

Resources may be natural or human. Those provided by the nature are natural resources ie. soil, climate, air, solar energy, water, natural vegetation, and animals, etc.

Human resources, on the other hand are provided by man. Man is himself a resource due to his positive qualities such as intelligence, skill, knowledge, culture, the physical and mental energy of man, etc. All there are human resources. Economic and commercial development of a country is a net result of interaction between natural and human resources. Thus, soil is a natural resource, but man applies his skills knowledge and experience to cultivate it, and obtains a variety of crops and raw materials from it. Similarly, rivers are used for transportation, irrigation and generation of hydro power. Mineral resources are used in many ways, plants and animals as well. Thus, economic development takes place as a result of interaction between natural and human resources. Further, resources of the world have been classified into Flow and Fund Resource.

On the Basis of Composition

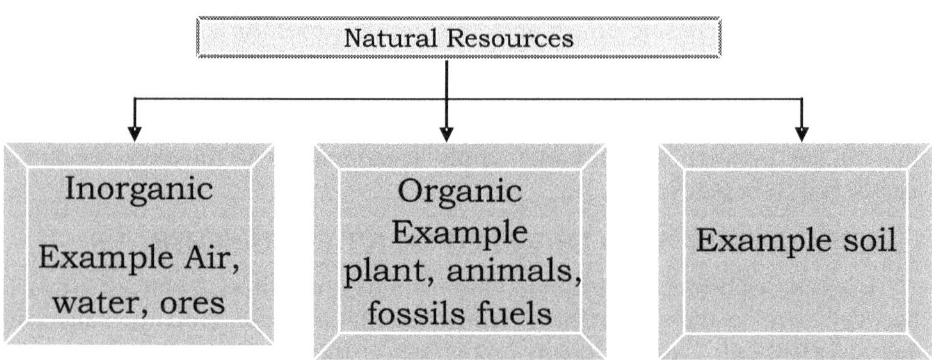

Natural Resources are the substances which ensure the material prosperity of the nations. Resources play a vital role in socio-economic development of each country in different parts of the world, as they are the main instruments of power and wealth. They affect man's destiny in war and peace alike (Zimmerman, 1957) (Hunker, 1964) and their proper use influences the health, security, economy and well-being of the nations (Kennedy, 1961).

On the Basis of Ownership

Resources are most commonly classified on the basis of their abundance and occurrences.

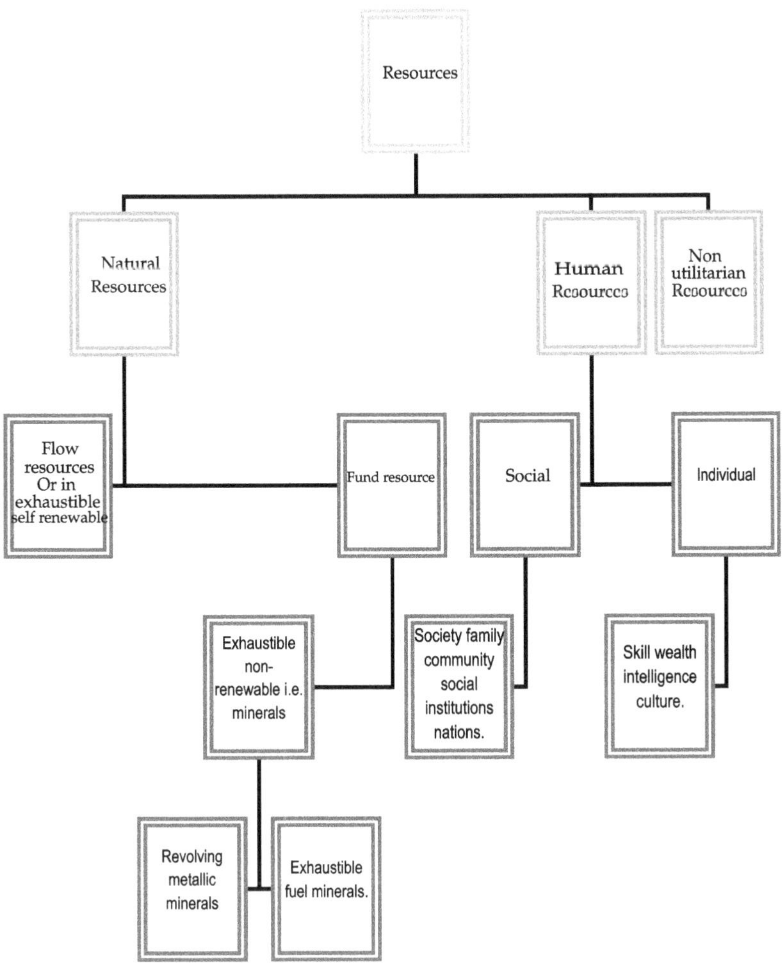

Source: Rashmi Desai (2007) Geography of Resources.

Natural resources are either Flow Resources or Fund Resources. When the supply of a natural resource is permanent, it's called a Flow resource. For example: while breathing, all animals and man absorb oxygen from the air and give out carbon dioxide which helps plant. Society in which photosynthesis is present helps in discharging oxygen. Thus the proportion of these gases in the atmosphere always remains constant. Air is, therefore, a flow or inexhaustible or renewable resource. Water supply is constantly maintained on the earth surface due to a perfect hydrological cycle.

Fund resources, on the other hand, are those which are fixed and limited in quantity. They do not renew themselves through automatic, natural process like flow resources as minerals belong to this category. They exist in fixed quantities in different parts of the world. Once extracted, they cannot be replaced. Thus, coal, petroleum, iron copper, manganese, gold, silver, bauxite and a horde of metals are fund resources.

Fund resources may be classified into exhaustible and revolving resources. Fuel resources, like coal, petroleum and natural gas, can be used only once. They are burnt and destroyed forever in the process of giving energy, and are therefore, exhaustible. Most minerals, on the other hand, are revolving resources. They do not get destroyed completely with one use, but can be reused by further processing, i.e., metal, equipments.

Human resources, i.e., the quantity and quality of man power or human resources is determined by the population that is characterized by number, density, age group, sex ratio, standard of living. Rate of literacy, technical skills, etc. These characteristics will decide the king of human resources that are available for production as well as consumption of materials and products. The over all economic development of a region depends upon its manpower for finally, it is man who decides to exploit, produce and consume.

1.5 FISHERY RESOURCE

Fish has become an important source of food for many centuries. Even today it is also the prime source of food substances in many countries. Earlier fish capturing was practiced by men with traditional methods and by applying their skills. Today the modern means of technology are at their disposal. As a result, the modernization has been adopted along with traditional methods. The mechanization adopted in marine fishing and traditional methods are practiced not only in aquaculture but also in marine fishing activities. The high level mechanization is found in the technically advanced countries, and vice versa.

Fishing is one of the extractive occupations of man. It is older than agriculture. Fish provides the easiest available food for man. However, fishing is unusually difficult and does not provide a livelihood for many people. It hardly absorbs less than 2% of the worlds labor force. In fishing, no cultivation of land is involved, and there is no waiting for maturity of crops as in agriculture. Fish provides an inexhaustible source of food supply owing to its incredible capacity for multiplying. Thus, marine life may have a very important role to play in the food of man in future, when the pressure of population increases beyond the capacity of soil to produce food grains.

All marine life lives on plankton. Vegetable plankton grows on the ocean bed where it is sufficiently shallow for sunshine and oxygen to reach it. It grows in abundance in shallow waters particularly in continental shelves and fishing banks. Animal plantation lies on vegetable plantation and is, therefore, abundant in shallow waters. The main food of fish consists of vegetables and animal plantation. The most important fishing grounds of the world are, therefore, found in depth of less than about five hundred feet. The one thousand feet line limits the depth of practically all fishing ground.

1.6 HISTORY OF FISHERIES

Fishery activity has its own historical background, and now-a-days it is also one of the main dominant determinants from the economic point of view in the coastal area, in particular, and source of water resource in general.

Fisheries and agricultural farming have evolved rather parallel in the history of human civilization. Interest in fish eating dates back to the dawn of the history. It is believed that hunting of fish was not uncommon in prehistoric times. At the developing sites near a river or lake of cave dwellers of the late Old Stone Age (40,000 BC) heaps of refuse of shellfish and sea fish have been found. The great distance by which these sites were separated from the sea points to some primitive way of fish preservation (like sun drying and smoke drying over the fire wood) practiced by these ancient dwellers to keep the food in edible of salmon smoking practices. Bronze Age (3500 BC) was the time when salting of fish started. Trading in dried fish, of course, was in vogue with ancient civilizations of Egypt, Mesopotamia and Indus Valley. Many cities were christened according to their fishing activities, viz. Sidon (meaning fishermen's town) in the Mediterranean. Among the Greeks and Romans, fish became an important food of the rich and the poor. Iron Age (1000 BC) saw the great trades in dried, smoked and salted fish in Greece. The time 400BC to 450AD was the period of emergence of highly organized fisheries in Roman Empire. In the Middle Age (500AD to 1500AD) fishing activities was more pronounced in the Atlantic, and herring and similar other fisheries gained importance. The empirical findings of the preservation methods, like salting, drying, smoking and potting in vinegar, of the medieval times have come down to the present times practically unchanged in their utility. It was not until the beginning of the 18th century that preservation in ice was started. The Chinese were the first to use ice in preserving the fish during transport. Ice preservation of fish was introduced in England about 1786. 19th century brought the time of scientific consideration, application of science to agriculture by the middle of the 19th century allowed agriculture to far outplace the fisheries industry. However, when application of science of fishing industry was made in the recent times, the position began reversing. Off-shore fishing, deep-sea fishing and fish culture hold enormous promise. The interest in the nutritive value of fish was first keenly taken where there occurred a sudden meat scarcity in the post World War 1 period. In twenties and thirties, the discovery of vitamins A and D placed fish in esteemed position because of the revelation that fish are generally rich in them.

1.7 TYPES OF FISHERIES

The major fisheries of the world fall into three categories. Their details are as follows;

i) Fresh water fisheries,

ii) Shore fisheries, and

iii) Deep sea fisheries.

i) Fresh Water Fisheries

Nearly two-thirds of the fresh water fish is caught in Eastern and Southern Asia. Many of these are raised in shallow ponds and flooded fields. Technically this type of activity is agriculture rather than fishing where fish is not caught from waterways and water bodies but is domesticated. Fresh water fish is more common in the tropics and middle latitudes. The largest supply of such fish comes from the shallow rivers and lakes. However, river fishing is generally done on a small scale by people inhabiting the river banks with primitive tools and equipment. Fish is caught for local consumption in small quantities. Fresh water fishing is important in countries having large dimensions where the interior parts are far off from the sea. Rivers, lakes and ponds provide the major supply of fish in such countries.

ii) Shore Fisheries

These are salt water fisheries as against their fresh water counterparts. Almost every country with a sea coast develops shore fisheries in the continental shelves. The extent of the shelf, the availability of plankton and the existence of ports determine the development of shore fisheries. Shore fishing is done on large scale by means of trawlers which are large fishing boats equipped with modern fishing tools.

iii) Deep Sea Fisheries

There fisheries are located in the shallow parts of the oceans away from the coast. These are known as fishing banks. Deep sea fishing involves the use of large fishing vessels equipped with most modern machinery and tools. Fish is caught in large quantities and is processed into a number of products on the ship itself. The fishing boats remain on high seas throughout the fishing season. Deep sea fishing, therefore, is more important for fish products rather than fish itself.

Fisheries can also be classified into tropical and temperate fisheries.

a) Tropical Fisheries

Tropical fisheries are much less developed than temperate fisheries because of a large number of difficulties involved in commercial fishing in tropical seas. The reasons for non-development of fishing in the tropics are as follows:

i) The sea water in the tropics has low plankton content, because of high temperatures of the waters which does not favour the growth of plankton, the major fish food.

ii) Fishing banks are conspicuous by their absence in the tropics. Tropical waters are quite deep except around Indonesia. Absence of large fishing banks is perhaps the major reason for non-development of tropical fisheries.

iii) Narrow continental shelves of tropical lands also cause a similar difficulty. The sea near the coasts of many tropical countries is quite deep.

iv) Nature has been more bountiful to tropical lands in respect to high temperatures, good rainfall, fertile soils, rich animal life and large forest resources. Alternative occupations like agriculture are thus more developed, and fishing is neglected.

v) High temperatures in tropics make fish preservation more difficult and costly.

vi) The coastlines of most tropical countries are unbroken, so that there is a paucity of good fish harbors.

vii) Tropical waters also contain numerous species of fish, but the quantity of each is limited. This makes specialization difficult. Fishing of particular species on a large scale, thus, becomes difficult.

viii) Tropical waters contain many non-edible and even poisonous varieties of fish.

ix) Ocean waters in the tropical regions are often disturbed by strong trade winds, monsoons and cyclones. This makes fishing difficult and risky, especially in certain seasons.

x) Tropical countries also suffer from lack of technical development, so that fishing tools and equipments are wide and primitive. Mechanized fishing trawlers are rarely used.

xi) Lack of suitable transportation and banking facilities has also retarded the growth of this industry in the tropics.

xii) Social and religious prejudices against eating meat and fish have also been responsible to some extent.

b) Temperate Fisheries

The temperate regions of the world have developed fishing as a major human occupation. The cool waters of the temperate seas provide fewer varieties of fish, but each variety is found in unlimited quantities. The fish move together in large shoals which are at times miles in extent. Each shoal contains one variety with the result that fishing becomes much easier and profitable in these regions.

Factors which favored these fisheries are as follows.

i) High Plankton Content

The sea water in temperate regions has high plankton content, which is the main food of fish. Low temperature favors the formation of plankton. Large quantities of plankton are also brought by the ocean currents like the Gulf Stream and the Kuro Siwo flowing from the tropics towards the temperate regions of the oceans.

ii) Large fishing banks

The most extensive fishing banks of the world are situated in the temperate regions. The Grand banks off Newfoundland in the North Atlantic, the Dogger banks in the North sea, the English channel and the sea of Japan are all situated in the temperate latitudes. These extension areas of shallow waters abound in fish.

Cool Waters

The large species of fish are generally found in cool waters of temperate areas, as against the small varieties of the tropics.

iii) Ocean Currents

Large shoals of fish move along with the ocean currents. In the North Atlantic, fish follows the route of the Gulf Stream and is caught near Europe. In the North Pacific, the Kuro Siwo brings large quantities of fish near Japan where it is caught.

Some of the ocean currents also form large fishing banks by depositing silt, mud and rocks on the sea bed.

iv) Fresh Water Rivers

The rivers of Western Europe bring large quantities of nitrogen and minerals and deposit them in the sea. Nitrogen helps the formation of plankton, and attracts fish. Some species of fish also move into river mouths for spawning and are caught there.

v) Broken Coastline

The broken coastline of Europe, North America and Japan provides many sheltered harbors with quiet waters where fishing is safer than the open seas.

vi) Fewer Species

The cool water of the temperate zone provides fewer species of fish, but each fish is found in unlimited quantities. The fish moves together in large shoals which, at times, are miles in extent. Each shoal generally contains one variety, which facilitates the catch considerably.

vii) Extensive Continental Shelves

The continental shelves near the coasts of Western Europe, Eastern U.S.A and Canada, and Japan are quite extensive. Fish abounds in the shallow waters of the continental shelves as in fishing banks.

viii) Low Agricultural Productivity

The low temperatures of temperate regions are not suitable for the cultivation of many food grains. The land available for agriculture is also highly limited in Japan and Western Europe. Agricultural productivity is, therefore, low while population is dense. Fish serves as a supplementary food in these regions.

ix) Large Population

Fishing is a labour intensive industry. A large labour supply is necessary not only for catching fish, but also for cutting, cleaning, salting, smoking, pickling, drying and canning of fish. Abundant labour is available in countries of Western Europe, China and Japan to carry out these processes. The large population also serves as a market for fish consumption.

x) Fish Preservation

In the cool climate of temperate lands, fish preservation is relatively easier than in higher temperatures of the tropics where refrigeration and cold storage increases the cost of fishing.

xi) Temperate Forests

Temperate forest such as those of Scandinavia and Canada provide both soft and hard wood for construction of fishing boats, barrels and casks, as well as pitch for water-proofing. These were once decisive factors in the development of fishing in these countries.

1.8 IMPORTANCE OF FISHERIES

Fishing is one of the important extractive occupations of mankind which is older than agriculture. The term fishery does not only mean taking away fish from water, but relates to the capture of all aquatic life and also other methods of catching all types of marine animals, legally termed as fish. Fishing is one of the important economic activities of the world as it provides a large number of employment opportunities not only in the process of catching but also in other ancillary and subsidiary related activities. Fish is the easiest available food for human beings.

In fishing, unlike agriculture, no cultivation of land is involved and there is also no waiting for maturity of the crops as in the case of agriculture. Fish provides inexhaustible sources of food supply due to its incredible capacity of multiplicity. Thus marine life may have a more important part to play as the food for man in the future when the pressure of the population increases beyond the capacity of supply of food grains.

The role of fishing has changed to a considerable level over the number of years. In the primitive time, fishing was mainly for consumption, but today in the modern world fishing is vibrant economic activity. The traditional method of fishing, too, has been replaced by modern scientific methods of fishing. Fishing is an activity, which is carried on in all the parts of the worlds, and the use of fish, too is varied.

The fishing activity provides employment either directly or indirectly to many people of Goa, and also contributes significantly to the total economic growth of the state. It gives boost to the canning industries, ice factories, boat making, basket making, etc. A couple of oil extraction as well as fish meals manufacturing units are also set up in the state. The fisheries also help in earning foreign exchange. Because of all these reasons, the fishing sector is developing rapidly.

Fish are vital source of food supply in countries, like Japan, Norway, Sweden, Iceland and Greenland where the land is bleak or mountainous and agriculture cannot be easily developed, and fish is also caught and processed to produce variety of products, i.e., lubricants, fertilizers or cosmetics, etc.

In addition to taste appeal and great variety, fish desires inclusion in the family diet on the basis of actual food value. Protein, fat, minerals and vitamins are the food elements in fishing products. In fact, about three fourths of the total fish catch is used for food whether fresh or processed.

Fish are rich in animal protein, the substance that replaces worn out body tissue and promotes growth. An average serving of fish or shell fish will supply the animal protein necessary for properly balancing the less efficient cereal and vegetable proteins in the daily diet. Fish are also valuable because of their mineral content, i.e., iron, copper, manganese, iodine, calcium, phosphorus, all essential to human well beings are a few of the minerals that have been accumulating in the sea for thousand of years. These minerals enter in to the bodies of small sea animals liver oils have long been known to be rich source of vitamins AD.

In addition to using fish as human food, large quantities of shell fish and fish are used in the manufacture of products. An important portion of the world catch is utilized directly in the production of fish meal, oil and other products, and great quantities of waste from canning, filtering and otherwise preparing fishery products of human food are also used in the production of by products. Whereas large quantities of fish meal were formerly used as fertilizer, nearly the entire production is now used for animal food. The meal provides protein of high biological value, and other elements makes it valuable in relation to poultry, pigs and young cattle.

Fish, whale and other marine animal oil have a large number of individual uses. They are used in the treating of lather and in the manufacture of soap, paint, printing ink, linoleum, oil cloth, lubricants and greases and many other products. Large quantities are also used in the manufacture of margarine. Certain fish body oil have sufficiently high vitamin A and D potency to make them valuable as feeling oils for poultry. Fish liver oils are an important source of vitamin A and D for over 100 years. But only recently has it been known that the livers of such species as sharks halibut, sword fish, sable fish ling cod, tuna and other are even richer source of vitamins A and D.

Economic Significance of the Sea

In olden days, the sea was considered to be a barren place, but with the advancement of civilization man gradually realized the importance of sea as a store house of food, i.e., fisheries, energy and various other necessities. The following are the important aspects of the ocean and seas of the world.

1. Sea as a source of food,
2. Sea as a source of minerals,
3. Sea as a source of energy.

1. Sea as a Source of Food

Sea is an important element of the human food supply. Fish accounts for nearly one fifths of all animals' protein in the human diet, and around 1 billion people rely on fish as their primary protein source. Indeed, production of fish is far greater than combined global production of poultry, beef or pork.

Traditionally, fishing is one of the most important economic activities in countries like Russia, Norway, Denmark, Iceland, Spain, U.K, etc., in Europe, U.S.A and Canada in North America, Peru Chili in South America, and China, India and Korea in Asian countries (continents). Among these continents, per capita food consumption is highest in North America.

2. Sea as a Source of Minerals

The sea can be considered as a store house of minerals on the earth.

a) Sea Water

According to the available estimates, one cubic km of sea water contains 120 million ton of salt, 6 million tones of magnesium salt, 20 tones of gold and 35 tones of silver. So 450 million cubic km ocean water can provide all over minerals requirements, if properly utilized. Besides these minerals, several nodule are available from sea floor. Most important mineral source is manganese and other minerals. The important nodules are copper and nickel nodules.

b) Sea Sands

Sea sands are another important source of minerals. Monoxide limonite, zircon are now being extracted from sea sand.

c) Iodine

Sea weeds provide ample iodine which gradually extracted for human consumption.

d) Petroleum

Among all the inanimate resources, perhaps, the most popular and economically important is crude oil exploration from sea floor. Sea beds have become an important source of oil in the present world to meet growing needs of the population.

3. Sea as a Source of Energy

It has been estimated that the sea can be considered as one of the greatest sources of potential energy. If properly tapped, it could provide more than 15% of world energy demand from tidal power and ocean thermal energy conversion

a) Tidal Power

Tides contain enormous kinetic energy. Its occurrence, power, strength and durability is highly predictable, so sustainable development of hydro- electric generation through turbines is feasible and is being practiced in some countries like Alaskan bay in U.S.A San José in Argentina bays of Cambay and Ketch in India.

b) Ocean Thermal Energy Conversion

Sun rays cannot penetrate much down the ocean water. So upper most layer of water receives heat and remains warm. Lower layer are comparatively cooler. So using this temperature difference or thermal gradient, conversion to eclectic energy is feasible. This easy unending sustainable energy may be harnessed. In warm areas, average temperature difference between the first two layers is around 25o C. Ocean thermal conversion plant can successfully tap this temperature and convert it in to mechanical and electrical energy at a low cost.

c) Ocean Waves

The incessant movement of waves possesses tremendous motive force which in turn can be converted into energy. Water, air operated turbines can convert this force in to different forms of energy.

d) Bio Conversion

This is another important technique used for energy production. Bio conversion of energy does not require any kind of machine operation. Sea weeds grow abundantly on the ocean floor, and possess unique quality of energy production conversion. Sunlight, after photosynthesis activities, is converted into bio-mass which can be successfully used as fuel to produce energy. The vast ocean floor can be used for artificial cultivation of sea weeds for viable commercial production of fuel and electricity. Different varieties of sea weeds macro cystis laminarta, eklonia together known as kelp may be cultivated.

The seedlings of this kelp are again re-cultivated in marine condition in a large scale. These cultivated sea weeds are able to produce ample methane gas which can be converted in to fuel. Besides fuel, food items and fertilizers can be produced from sea kelp.

1.9 GLOBAL SCENARIO OF FISHERIES

More than 200 million people in the world are involved in fishery activities which contribute 3% of working population. The 80% of the working population lives in the vast costal belts in the world. They have been involved in the subsidiary related to the fishery resources in different forms. The small scale household activities, process and preparing the fish for sale transaction, etc. The Global fisheries production in 2001 was worth US$ 78 billon at the first sale aquaculture (fish farming) was worth US$ 60 billion in 2002.

Table-1.1 : Production of Fishery Resource on a Global Level

Production	2002	2003	2004	2005	2006
INLAND					
Capture	8.7	9.0	8.9	9.7	10.1
Aquaculture	24.0	25.5	27.8	29.6	31.6
Total Inland	32.7	34.4	36.7	39.3	41.7
MARINE					
Capture	84.5	81.5	85.7	84.5	81.9
Aquaculture	16.4	17.2	18.1	18.9	20.1
Total Marine	**100.9**	**98.7**	**103.8**	**103.4**	**102.0**

Source: The state of world fisheries & aquaculture, 2008 (F.A.O. 2008)

World fish production as per latest statistics released by the F.A.O rose from 90.05 million tones in 1991 to 98.11 million in 1992. India emerged in 1998-1999 as the largest fish producer in the world. China is now at the top amongst fishing nations during 2006-2007.

The following table displays the production of inland and marine fisheries in the world over the years. The inland fisheries mainly consist of capture and aquaculture, produced around 41.7 million tons in 2006 as compared to 32.7 million tons in 2002. The trend shows gradual increase in the production of capture and aquaculture. Similarly, marine fisheries have grown 98.7 million in 2003 to 103.4 in the year 2005, but further marginally declined by 2006. It has been observed that marine prawns have grown from 16.4 million tons in 2002 to 20.1 million tons by 2006. This shows steady increase in marine aquaculture production over the years. Marine capture fisheries, seems to be constant in production till 2005, but declined marginally by 2006 at 81.9 million tons.

The following table shows the total fisheries and aquaculture production in the world.

Table-1.2 : World Fisheries and Aquaculture Production (Million Tones)

Year	Total Capture	Total Aquaculture	Total World Fish
2002	93.2	40.4	133.6
2003	90.5	42.7	133.2
2004	94.6	45.9	140.5
2005	94.2	48.5	142.7
2006	92.0	51.7	143.6

Source: Food Agricultural Organization (2008).

From the trend, it appears that the total world fisheries production has increased from 133.6 million tons in 2002 to 143.6 million tons in the year 2006, which shows great demand for fishery products, inclusive of capture and aquaculture.

The following table shows the global trend of human consumption and non food uses of Fishing Resources from years 2002-2006.

Table-1.3 : Trends of Utilization of Fisheries

Year	Human Consumption	Non food uses	Population in (billions)	Percapita
2002	100.7	32.9	6.3	16
2003	103.4	29.8	6.4	16.3
2004	104.5	36.0	6.4	16.2
2005	107.1	35.6	6.5	16.4
2006	110.4	33.3	6.6	16.7

Source: Food Agricultural Organization (2008).

The above table displays the pattern and trend of Human consumption of fishery products and the years in different parts of the world. It appears like Human desire for fish consumption has been increasing from 16 kg in the year 2002 to 16.7 kg in 2006-07. This per capita consumption of fish has been increasing as it is becoming an important item of human diet, which contributes 8.87% of the world fish catch of 143.6 million tones. By 2010 A.D. world fish production is expected at more than 150 million tones, of which it may be possible for India to contribute to the order of 10%. West Bengal achieved top most position in Inland Fish production during 1993-94. F.A.O estimated that there was a supply gap during the period 2000-2010 A.D. and 2020-25 are expected to be about 37 million tones and 63 million tones respectively.

1.10 NATIONAL SCENARIO OF FISHERIES

India's plentiful bounty of nature and biodiversity in regard to fish germplasm resources are amazing. The country is endowed with abundant inland and marine germplasm resources. While about 25,416 fish species are believed to inhabit water bodies all over the world, we know that about 2500 species are considered economically important. The judicious and rational exploitation of these renewable resources is of great importance, if we are to conserve this wealth for posterity.

India has a vast potential of fishery resources with 2.02 million square kilometers area of EEZ (Exclusive Economic Zone), about 45000 kilometers length of rivers and about 1,97,024 kilometers of canals, about 3.15 million hectares water has spread in the form of reservoirs and about one million hectares of coastal land awaits utilization through brackish water farming and number of protected water ways and covers along the 8118 kilometers coastline for remunerative Mari culture.

The following table depicts the status of Marine fishery Resources of India.

Table-1.4 :Status of Marine Fishery Resources of India

Sl. No.	Resources Position	Unit	Extent / No.
1.	Length of Coastline	Km	8,118
2.	Continental Shelf area	Lakh sq.km	5.30
3.	Exclusive Economic area	Million sq.km	2.02
4.	Rivers and Canals	Km	1,97,024
5.	Reservoirs	Million ha	3.15
6.	Ponds and tanks/area	Million ha	2.35
7.	Fresh water ponds/area	Million ha	5.40
8.	Brackish water ponds/area	Million ha	1.24
9.	Potential water ponds/area	Million ha	1.67
10.	Fishing Villages	Nos.	3937
11.	Estuaries	Million ha	0.29
12.	Marine resource potential	Million mt	3.90
13.	Present level of exploitation	Million mt	3.60
14.	In land resource potential	Million mt	4.50
15.	Present level of exploitation	Million mt	2.04

Source: Centre for Marine Fishery Research India (CMFRI), Cochin (2009).

Chief Fishing States

1. **Andhra Pradesh**: With a coastline of 960 kilometers in the Bay of Bengal and with 319600 hectares in inland waters is very rich in its fishery resources. The fish production of the state is estimated at 158000 metric tons of marine fish and 84,500 metric tons of inland fish actually.

   There are 13 fish farms at Ippur, Mopad, Sunkesula, Patha, Cuddapah, Hussain-Sagar, Rajendra Nagar, Dindi, Manair, Kolisagar, Nizamabad, Shanigram, Wyra and Warangal.

2. **Kerala:** The smallest maritime state in India with just over 590 kilometers of sea coast, contributes to a little over 25% of the total sea fish landed in India. Oil sardines (Sardinella longiceps) and the club mackerel (Rastrelliger Canagurta), the two most important shoaling fishes of the west coast, contribute to the major fishery of the coast, the third important constituent being prawns. The estimated landing of all fish in the state is about 225,717.4 m.tons of which 7932 m.tons are contributed by inland water fishes, the rest being sea fishes. This quantity is valued at about Rs. 6,77,43,393. 70% of the population in Kerala is fish eaters.

3. **Gujarat:** has a coastline of over 1649 kilometers extending from lakhpat in the north to Umbergaon in the south. It has a total fishable area of 120000 sq.kms., including the Gulf of Kutch and Gulf of Cambay.

4. **Maharashtra:** Maharashtra state has a sea board of more than 720 kilometers, extending from Zai (district Thana) in the north. The inland fisheries extend to Bhandara district in the east, a crow flight distance of over 800 kilometers. The most important varieties of sea fish are caught along the coastline: Saranga (white pomfret), Halwa (black pomfret), Ghol (Jew fish), Giant (threadfin), Rdavas (Indian salmon), Kuppa (tuna), Surmai (seer fish), Boi (mullet), Mushi (sharks), Wam (eels), etc.

   Catch of the fish from marine resources was estimated at 2,26,900 tones, and that from inland resources was estimated at 13,250 tones during the year 1968-69.

5. **Karnataka:** Karnataka state has 320 kilometer of coastline on the Arabian Sea. All along this board, there are many fishing villages, the most important of them being Karwar, Ankola, Kumta, Honnawar, Malpe, Udiyawar, etc. The total fish production on Karnataka coast is one-seventh of the total marine fish landing in India.

6. **Bengal:** Bengal is one of the most important fishing centres in India. The important varieties of fish caught along the coastline are eel, cat fish, phasa, hilsa, bhola, prawn, skate, etc. Ancillary industries like shark liner oil, gelatin, glue, fish meal, and their products: one scheme for production of shark liner oil, fish-meal, processed fish and utilization of other fish by-products is operated at Contai coast.

7. **Tamil Nadu:** It has a coastline of 997,47kms, with 234, 00sq, km of inshore water and a continental shelf of about 7,8000sq.km and about 780912 hectares of inland waters. There are six fishery training centres at Tuticoein, Nagapattanam, Mandapam, Colachel and Madras.

8. **Orissa:** Orissa is also one of the most important fishing states in India. There are many inland estuarine and marine fishing centres. The important varieties of fish are caught along the coastline: pomfret, catfish, sharks, rays, skates, mackerel, silver bellies, etc.

India is the third largest producer of fish next only to China and Peru, and it ranks second in the aquaculture production farmed inland fish in the world. The following table depicts the trend of Fishery production and share of export in international trade from India.

Table-1.5 :Statewise Scenario of Marine Fishery Resources

State/Union Territory	Approximate Length of Court Line (km)	Continental SHELF (000′ km)	No. of Landing Centres	No. of Fishing Villages
Andhra Pradesh	974	33	508	508
Goa	104	101	88	72
Gujarat	1600	184	286	851
Karnataka	300	27	29	221
Kerala	590	40	226	222
Maharashtra	720	112	184	395
Orissa	480	26	63	329
Tamil Nadu	1076	41	362	556
West Bengal	158	17	47	652
Andaman & Nicobar Islands	1912	35	57	45
Daman & Diu	132	4	11	10
Pondicherry	45	1	28	45
Total	**8118**	**530**	**1896**	**3937**

Source: Centre for Marine Fisheries Research Institute (2009).

Table-1.6 :Trend of Fishery Production and Share of Export in International Trade from India

Year	Marine Fish	Inland Fish	Total Fish Production	Export Value (Rs. in crore)	Export Qty. Qty.	Export as % as %
1950-51	5.34	2.18	7.52	2.00	.020	2.66
1960-61	8.80	2.80	11.60	4.00	0.19	1.64
1970-71	10.86	6.70	17.56	35.00	0.36	2.05
1980-81	15.55	8.87	24.42	235.00	0.75	3.07
1990-91	23.00	15.36	38.36	893.37	1.39	3.62
2001-01	29.11	28.45	56.56	6296.00	5.03	8.89
2006-07	31.00	38.00	69.00	8363.53	6.12	8.87

Source: Department of Animal Husbandry, Dairying & Fisheries, Ministry of Agriculture, New Delhi (2008).

It has been observed that India has become the 16th largest exporter of Marine products in the world, and the 9th largest exporter of Marine products in the Asia. Fisheries sector of India accounts for 5 percent of the global fish production and 2.5 percent of the global fish trade. The USA and Japan are the two largest importers of Indian Marine products with a share of nearly 32.50 percent in 2006-07. The fisheries sector contributes Rs. 19.555 crores to India's National Income which is 1.4 percent of the GIP and 4.7 percent of India's agricultural GDP. The per capita availability of fish in the country has increased from 3 kg to 9 kg. It showed a majority of India's fish production around 90 percent is used for domestic consumption, and hardly 8.87 percent goes for exports. There is still an immense export potential for Indian fish products in the ocean seas markets. Out of the total Indian exports, the share of sea food export is 3.32 percent now. However, it is the 3rd largest contributor to the net foreign exchange earned by the country.

1.11 MARKETING AND ITS IMPORTANCE

"Society can only exist when only a large number of people want something a few people have. It is necessary for both groups to be mutually aware of this need".

Marketing is the so basic that it cannot be considered as a separate function. It is the whole business seen from the point of view of its final result, that is from the customer point of view, business success in not determined by the producer but by the customer – Peter Drucker (Philip Cotler, 1997).

Marketing may be defined as "the performance of business activities that direct the flow of goods from producer to consumer or users". It may be said that marketing includes all those activities which effect changes in the ownership and possession of goods and services. It may be noted that it is closely related to that part of economics which deals with the creation of time, place and possession and utility, and helps in satisfying human wants by the exchange of goods and services for some valuable consideration. Sherlekar (1986).

Marketing is the process of discovering and translating consumer needs and wants into product and service specifications, creating a demand for these products and services and then, in turn, expanding this demand. In fact, the concept of markets finally brings useful circle to the concept of marketing. Marketing means working with markets, which, in turn, means attempting to actualize potential exchanges for the purpose of satisfying human needs and wants.

"Marketing is concerned with the people and the activities involved in the flow of goods and services from producers to consumers". There are three objectives of marketing: they are, to fulfill needs, to satisfy wants and to create new desires.

"Marketing is the management function which organizes and directs all those business activities involved in assessing and converting consumers purchasing power into effective demand for a specific product or service and in moving the product or service to the final consumer or user so as to achieve the profit target or other objectives set".

At a glance the functions of marketing are:

All of these are marketing functions involved in marketing of goods and services.

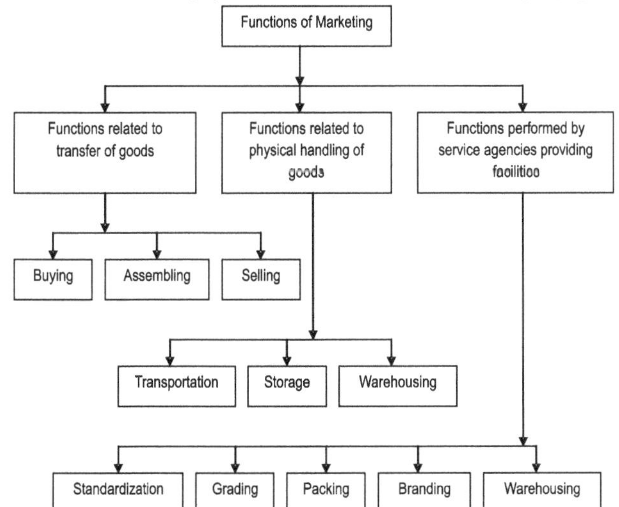

Source: Sherlekar (1986) Marketing Management.

1.12 MARKETING OF FISHERIES

In earlier days, the term marketing of fish meant buying and selling of fish at the leading centers. After the Second World War, marketing of fish has taken a new role in business activity. The fisheries have become highly industrialized in all fishing nations. The new fishing techniques have been adopted to sell more fish. The modern fish marketing systems lays emphasis on meeting the existing demand of fish, besides tapping the potential demand in the important markets. The marketing of any procedure mainly depends upon the availability, consumption and demand. In Goa, traditional system of fish marketing is adopted. The fish marketing is normally done at the collection centers, which are mainly situated in the area of fish landing.

Fresh fish may be sold daily in fishing ports, located near fishing areas. However, fish and fish products to be sold in distant markets must first be processed to prevent spoilage. Many fishing crews sell their catches to process at auctions after fishing trips. The price which a catch commands depends on the supply of fish at the market and demand for it. A fishing crew does not know in advance what a harvest it will earn if it is sold at all. Sometimes, the processor, place orders with the fisherman before fishing trips. At the same time, both sides agree on the price to be paid for the catch. Sometimes, fish may not be consistently demanded by customers due to its peculiar smell. It is an act of fishery technologist to prepare the odorless fish products for wider consumption. Thus, the fish marketing objective should not only be catching and selling of fish but, should have a wider scope for exploitation, production, distribution, preservation, packaging and transportation, etc., of fish, in addition to actual sales by reducing middleman.

1.13 MARKETING GEOGRAPHY

Introduction

The history of German Civilization reveals the fact that man relied very heavily upon the nature for the satisfaction of all his wants. With the advent of human civilization gradually, he began to free himself from the clutches of nature. To begin with, the human needs were very limited, and man was contented with whatever he had. As days advanced, human wants increased there by the scarcity began to exist. This paved the way for exchange of goods and services between different societies. Accordingly, the 'barter system' came into existence gradually. As a result, the surplus and deficit areas were developed in the human society. As a measure to maintain a balance between the surplus and deficit areas of particular commodity (commodities), a specific market system developed (evolved). In due course of time, the human wants increased remarkably, giving rise to an acute shortage of commodities, and requested into exchange of goods and services between the societies through a media called 'money'. Presently, money has become the major medium of exchange to facilitate exchange of goods and services between the surplus and deficit areas. The sufficiency and deficiency have paved the way for the emergence of markets, and these places have become the centres of exchange. The sphere of this exchange, which was confined to smaller areas, has covered the entire earth today. As a result, markets are integrated globally, and there one can speak of global markets and global exchange, etc.

Concept of Marketing

Marketing Geography is a recent origin of the late 1950's. It has attracted the Geographers and planners, and accordingly moved in the research activities. Marketing Geographers are interested in trading areas, which are complex in real manner. Therefore, they seek to provide scientific knowledge and scientific assistance to business men in evaluating market area potentials.

Marketing is an integral part of economic geography. There is no doubt that marketing is a part of man's economic activity. The need of marketing geography was first pointed out by William Applebaum (1954), an American Geographer and argued that the study had been neglected by geographer upto that time, inspite of large sections of the working population engaged in marketing functions, and a large part of urban landscape was devoted to structures of wholesale and retail trade. The complex of channels of distribution, from the producing area to consuming area are the faucal theme of Marketing Geography. The main root if marketing geography is the market place.

The Term "MARKET" has been derived from the latin word "Mercatus", which means the place or method of contact between buyers and sellers. Applebaum (1954), who is widely regarded as the chief architect of Marketing Geography as a distinct field of study in the United States, defined Marketing Geography as a branch of knowledge "concerned with the delimitation and measurements of markets and with the channels of distributions through which goods move from producer to consumer."

A market, to an economist, is a specific institution with rules of its own, upon which it builds up powerful analytical structure. (Bromley 1971 and 1976, Arunachalam 1983)

Geographically, Market is a well defined area, varying from a small village market to a metropolitan centre markets, and are inter-related with different geographical aspects, which perform its functions deeply involved with the behavioral approach of different market merchants.

Since the market (market) is a place of exchange, it indicates the process of functions of the market. Such a process of functions could be named as marketing.

Abbott (1958) define that "Marketing includes all the activities involved in the flow of goods and services, from production to consumption".

- Yeung (1973) opined that "Marketing is a common day to day activity which involves the exchange of goods and services from producers to consumers".
- Hodder (1965a) has described market as "An authorized public gathering of buyers and sellers of commodities meeting at approved place at regular intervals".
- Berry (1967) has described that "Marketing Geography put the retail interests of the geographers into practice within metropolitan areas in the service of private business enterprises".

According to Davies (1977), it is mainly concerned with the tertiary economic activities, and particularly the distributive trade.

Belshaw (1963) opined that the "Markets are sites with Social, Economic, Cultural and other reference marks, where there are a number of buyers and sellers and sellers, where the price offered and paid by each is affected by the decisions of the others. In common usage, the elements of a market are buyers, sellers, merchandise and a fixed place and time for exchange. Each market has two functions to perform. They are: (1) the distribution of local products, and (2) the exchange of rural surplus for urban goods.

The functions of marketing is an integral part and play a vital role in the modern capitalist system. It is a major factor in the differential growth of cities and in the changing Economic status of rural areas. Some markets are performing with specific goods and services, are highly specialized with a relatively small number of users, for other goods and services, the demand is wide spread. The market and market functions have become so effective with the efficient channels of distributions when the surplus and deficit areas have been linked by the market. It is observed that there is a considerable movement of men and materials from one part to another part, which could be regarded as spatial mobility. Market centres perform various functions not only for the rural and tribal, but also for the urban people.

The cost of goods is affected by the distance travelled to reach the place or destiny. The distance travelled and the mode of transport have a significant effect on the determination of prices of commodities. Thus the market centres play a vital role in the socio-economic development the region or country.

Market Place

Market Place is one of the basic components in marketing geography, and plays a significant role in development of marketing activities in dynamic manner, arranging different spaces and merging into a place for exchange of commodities and services. As Hodder (1965a & 1965b) defines, "Market place is an authorized public gathering of buyers and sellers at a fixed place during stipulated hours, in which the exchange of commodities takes place".

- Berry (1967) made a pioneer attempt to put marketing geography in a theoretical frame for continuing for further research. He published a book titled "Geography of market centres and retail distribution which is first of its kind in the history of the development of marketing geography and laid principles concerning the spatial distribution, organization and behaviour characters of markets of the United States and money developing economies also.
- Dixit (1979) defines that a market place is a geographical space, which attracts the buyers and the sellers to exchange their required goods and services in a given time. Market place is an area of demand for exchange of the goods and services within clearly defined geographical limits.
- Hugar (1987) describes that the market place exchange is geographical space is attracted by its functions of exchange of goods and services in a specific day in a given interval.
- Mulimani (2002) view on market is that, market place is a public assembly where buyers and sellers are meeting to each other and exchange required goods and services in a specific day in a given interval.

The market can be categorized on the basis of time duration. If the market function is observed on all the days of a week it is a regular or daily market, and if it is observed once in a week it is called weekly market. When the transaction of agriculture produce is controlled by the government it is said to be regulated market. However, in the state of Goa all the three categories of markets, namely, daily, weekly and regulated are observed.

Periodic Markets

The periodic markets are the age-old traditional markets and are found in Urban as well as in rural areas. Periodic markets are basically places of exchange of goods and services to the residents of a given geographical area at a fixed interval of time (Hugar, 1982 & 1984).

As Bromley (1975) opines the periodic markets are authorized gatherings of buyers and sellers of commodities, meeting at an approved place at least once a week but not as frequently as daily.

Non-Periodic Markets are those which function daily or regularly and also under the control of government in order to manage the wholesale transactions of agricultural produce, that is, APMC all District level. Mobile traders as well as permanent village shops fall into this category to cater to the needs of growing rural and urban population. The main function of the market is to provide better prices for goods and services to be exchanged.

1.14 REVIEW OF LITERATURE

The proposed study has made an attempt to review the literature pertaining to the marketing geography and Resource geography not only by the Indian scholars but also by the foreign scholars.

Dickenson was the first scholar who floated the concept of marketing in 1934, and studied markets and market area of East Angela. After that many years later, it was not used by any scholar.

William Applebaum (1954) was a noted scholar who is instrumental to motivate and to pursue the studies in the field of marketing geography and who gave it the relation of a distinct branch in the field of economic geography in the United States. His studies aimed at delimitation of market area and measurement of markets with various channels of distribution of goods from produce to consumers from the geographical point of view.

He initiated a group of scholars to pursue their study in the field of marketing geography in order to make it more applied in nature. This view has encouraged a number of scholars to carry out research effectively in the field of marketing geography. In view of these outstanding efforts, he was recognized as the chief architect of marketing geography. **Murphy (1961)** substantiated the same opinion of **Applebaum (1954)** and laid greater emphasis on the fact that marketing geography is one of the oldest branches of Geography and he laid a greater emphasis on study the metropolitan markets of the United States. **Hodder (1965), Berry (1967), Bromely (1974), Smith (1972)** and others were supporting pillars for the emergence of marketing geography.

Berry (1967) has opined that marketing geography has put the retail interests of geographers into practice within metropolitan areas in the service of private business enterprises, and of late, the term marketing geography is taken to describe that aspect of geography which is concerned with the tertiary economic activities, and, particularly, the distributive trade.

- **Bromely (1971)** has also done an extensive bibliography of market place studies, periodic markets, daily markets and fairs.
- **Carol A. Smith (1976)**, an anthropologist, carried out an exhaustive review of marketing geography in the light of central place theory by considering the various schools of thought. According to Christaller, the market centers are fixed in space and time, and he is well known for his central place theory much interested in understanding the various aspects settlements distributed in a given geographical area with population as a common factor. But the most basic element of his central place theory was the capacity to provide goods and services to an area and population larger than itself. Thus, he made it evident that the settlement as market places is the chief characteristics of his central place system (Christaller, 1933 & 1966).
- **Cleef (1937)** market centers vary in their nature, size, function and hierarchy, Courtney to urban markets. Rural markets are the primary centers of marketing, and provide a link between producers and consumers. Thus, they form a basic element of economic landscape of a region.

In India, a few attempts have been made by geographers to study the markets. As early as 1940, C.D. Deshpande and Krishnan have made pioneering efforts to study the market.

Bhat (1976), Mishra (1965) and **Tamaskar (1966)** have made outstanding efforts and studied the various aspects of marketing geography.

Alam (1963) concentrated on the markets from different points of view by considering both retail and wholesale markets in a metropolitan environment.

A new approach in modern marketing research has been adopted by considering the system of markets in different economies. In this regard Wanmali (1981) works on tribal economies, Dixits (1984) on Ganga Plain.

Hugar's (2000) study on thought proves economy and **Shrivastvas (1987)** study on Terai region economy are worth mentioning. Further in depth studies have been carried out by these scholars on physical or administrative levels in India. In Recent times, the marketing geography has been further streamlined and bifurcated into geography of retailing and wholesaling marketing.

RESOURCES

Resources are the materials which are found on the earth surface called endowments, i.e., soils, minerals, air, water, sunshine (solar energy) and natural vegetation, etc. Their usefulness to man and their capacity to satisfy a number of human wants makes them resources.

According to **Erich W. Zimmerman (1951a)**, "A man – less universe is void of resources : for resources are inseparable from man and his wants. They are the environment in the service of man". Availability for human use not mere physical presence, is the criterion of resources.

Erich W. Zimmermen (1951b) defines recourse as just a thing or a substance but the function that the thing or substance performs in order to satisfy human needs and wants.

Resources are the foundations of power and wealth, and they affect man's destiny in war and peace alike **(Zimmerman, 1957) (Hunker, 1964),** and their proper use influences the health, security, economy and well being of nations.

Ginsburg (1957) defines natural resources as "All the freely given material phenomenon of nature within the zone of man's activities plus the additional re-material quality of situation or location.

Harvey (1973) described "Resources as social and technological appraisal related to objects". The society oriented and relative value of natural resources is particularly prominent in an economic cum geographical considerations.

Oxford dictionary defines the resources "as a means of supply of some wants or deficiency, a stock or resources upon which one can draw whenever necessary".

FISHERIES – AS A RESOURCE

Susreta (600 BC) made ecological classification of fishes. **Kautilya (300 BC)** mentions in his Artha Shastra about fish culture in tanks and reservoirs. **King Ashoka (246 BC)** declared edible fishes as protected species during their spawning periods **(Ashoka's Pillar Edict V 246 BC)**

King Someshwara III, son of the king **Vikramaditya VI 1126-1138 AD,** Possessed immense knowledge of herbivores, carnivores and detrivores – feeding the fish.

Bloch (1785) published a book "Austandiche Fishe" Ichthyologie. The same was further extended by **Schneider (1801)** which contains many Indian Species. **Russel (1803)** described 200 species of fishes.

Hemitton made a description of 269 species of fishes from the Ganges and its tributaries in his pioneer work "Fishes of Ganges". In between **1828–1849 Cuvier and Valencienne** published a book entitled "Historic Naturellae des poissons". This provides remarkable information on Ichthyology.

Mac Clevland (1839) provided notable information and works on the study of Ichthyology in India.

Hora and Mishra (1920, 1959) carried out research on fish and fisheries, and provided authentic information for aquaculture.

John Kurien (1982) has studied "Technical change in fishing and its impact on fishermen in Kerala". He found that there was a considerable increase in output of fish in Kerala. He examined that the mechanized sector received the major share of the benefit of the increased production due to technological development in the fisheries sector.

Rao Subba's (1986) study "Economics of Fisherie's" reveals that the quantum of financial resources allotted to the fisheries sector in general and to specific schemes in particular is very limited. There is a lack of co-ordination among various agencies connected with the implementation of the different fisheries programmes to increase the fishery production.

Pervez De Souza (1988) in his study "Caught in a Net" carried out in Goa observed that mechanization of fisheries post liberation has increased the fishery output. In the process of increasing fish catch, the government has ignored the marine ecology and traditional fishermen.

Panniker and Sathiadas (1989), in their research paper "Marine fish marketing trend in Kerala", examined the fish marketing system prevailing in Kerala. Their investigation indicated that due to lack of infrastructure facilities, the supply of fish at the landing centers is highly inelastic, which often would result in disposal of fish at throw away prices at the time of heavy landings. The involvement of a number of middlemen in the marketing chain adversely affects the interest of both the fishermen and the consumer.

J. Vasant Kumar and K.R. Sundarvardharajan (1990), in their research paper, "Adoption of Scientific Technology by Trawler Operation (Operators) of Tamil Nadu", studied various aspects related to adoption of Scientific Technology by Trawler operators of Tamil Nadu. The survey or experiment was conducted in six maritime districts of the Tamil Nadu.

Sathiadas R. and Panikkar K.K.P. (1992), in their research paper "Share of fishermen and middlemen in consumer price : A study at Madras region", discussed the marketing margins and producers and middlemen's share in consumer's rupee for commercially important varieties of marine fish in Madras region of Tamil Nadu. They opined that by introducing co-operative marketing system involvement of too many intermediaries can be avoided to increase the efficiency of fish marketing system.

Ibrahim P. and Disilva S. (1994) have studied "Economics of mechanized boats and motorized crafts". They used various measures like catch per trip and effort, labour productivity, etc. Their investigation revealed that mechanized boats (Trawlers) neither affect a bigger output nor make a larger profit than the motorized crafts. This certainly calls for a shift in the mechanization policy of the government.

Silas (1995) study on "Conservation and Regulation in Marine Fisheries" reveals that fisheries management should aim at a sustainable long term economic utilization of the resources.

Sathiadas R. and Kumar Narayana (1994) studied "Price policy and fish marketing system in India". They opined that the growth of fish production and development of fishery sector is highly dependent on an efficient marketing system. They emphasized that modern fish marketing policy should envisage not only meeting the existing demand for fish, but also tapping the potential demand.

Kalia (1996), in his research paper entitled "Financing Fisheries Sector", discussed the fisheries scenario in the country and the credit aspects for specific projects. The NABARD has been providing grants from its research and development funds for conducting further operational research, standardization and commercialization of technologies.

Bhaumic and Saha (1998) in their study "Role of extension in arousing mass awareness and public participation in fish conservation movement", throw light on the role of extension mass awareness and public participation in fish conservation movement in India.

Sutton Muchel (1998), in his research paper "Harvesting market forces and consumer power in favour of sustainable fisheries", suggests that greater public involvement in the fishery management process must be encouraged. Marketed economic incentives must be created to promote sustainable fishing.

Immanuel Sheela and Srinath Krishna (2000) studied "Potential Techno-Economic Role of Women in Fisheries". The study revealed that women contributed a lot to fisheries sector. In coastal areas, women play an important role in fisheries, and in some parts of the world they are good navigators too.

Vijay Kumaran K. (2001), in his research paper "Social audit – an ideal method for neutralizing conflict situations in Aquaculture industry" made an attempt to study the activities of aquaculture industry having an impact on the immediate environment and to elaborate how social audit would help to neutralize the conflicting situation in aquaculture industry.

S. Venugopal (2005), in his book Aquaculture claims that information and consumer education programmes can play a vital role in expanding the demand for aquaculture products. He suggests that economists should be innovative in their research approaches to study consumer demand for fish and sea food, and also about the resource allocation and public policy, affecting aquaculture.

S. Subramanian and Neelam Komarpant (2003) in their research paper "Application of remote sensing techniques in increasing efficiency in Marine Resources", stated that remote sensing is an advanced technique that has been utilized in fisheries forecast in recent years. Fishes are known to react to changes in environmental conditions and congregate in areas where favourable conditions prevail. Parameters such as sea surface temperature (SST) and chlorophyll, besides other physicochemical and meteorological factors are used in identifying the potential fishing zones by using satellite sensing.

Dr. Ramchndra Bhatta (2003) in his research paper "Status of Marine Resources Exploitation and Management in Karnataka" examines. The marine negotiation and management regimes have not really prevented the over exploitation of fishery resources around the world, and India is not an exception. He analyzed the trend in fish production and other environmental changes affecting equity and sustainability of marine fish production in Karnataka. He highlighted that the prospects of improving the fisheries contribution to income and employment of state depends on the implementation of a set of regional and national fishery management policies which needs the immediate attention of the planners.

Aarthi Shridhar (2003) in her research paper "Conflicts and Related Issues in Marine Resources Conservation" observes that conservation strategies include habitat protection and species preservation. The Indian Coast has 5 marine protected areas (comprising marine sanctuaries and National parks declared under the wildlife protection Act 1972) some of which are pending final notification. Declaration and implementation of marine protected areas (MPA's) has followed conventional land (especially forestry) based management methods and process. This strategy of conservation has a reputation of disregarding and alienating local communities. In the complex social condition along the coast (Indian Coast) the fallouts of alienation are aggravated by differential resource areas patterns, social in equalities, imbalance and inappropriateness of technological applications and acute resource depiction. Former (legal) rights over the coastal lands, water regimes and ambiguous in many status leaving a sizable population without possession of formal records confirming their entitlements. The author examines marine conservation strategies in place in the country today and the implications that these measures (conservation measures) have for the livelihood of traditional and artisan fishing communities.

S. Balkrishnan (2008a), southern economist, in his research paper "Women Employment in Fisheries Industries: A Study", states that, Fishery Industries are the backbone of India. The emerging fishnet industries have the power to induce a social change in the rural areas of our country. The fishery industries are not only giving opportunities of employment to the rural folks, but also creating due recognition and status to them. They increase their income and cause changes in their social out look and attitudes. In the field of the study area, fishery industries are serving a lot for women employment social emancipation solidarity and socio-economic improvement of the community.

P.A. Koli (2008, Vol. 47) in his paper "Problems of Fisheries Co-operative – A Study" published in Southern Economist states that, the Indian Fisheries Industry, in general and fisheries co-operatives, in particular, is facing problems. The over fishing and over crowded boats must be strictly prohibited, otherwise, the famous tragedy of commons will soon be experienced.

Avinash Raikar (2003), in his research paper "Sustainable Development of Marine Resources in Goa what needs to be done?", opines as follows:

In recent years the composition of fish catch is also undergoing significant changes. The production of oceanic species, like shark, is increasing at the CAGR compound Annual Growth rate of 25 percent, whereas high value species like pravence has a negative CAGR. This had an impact on the exports of fish products and profitability of the mechanized segment of fishing. However, in the recent years, reduction in the harvest and experts of high value fish, like prowers, squids, lobster, etc., have led to increase in the experts "low" value fishes, like ribbon fish, mackerels, tune, etc., to the Asian and middle Eastern countries. This tread has seniors on the food security of 90 percent of the fish consuming popularity in Goa. The high demand would further encourage unsustainable fishing practices on the Goan coast and inland water systems. It is, therefore, essential to increase the production and availability of fish in Goa. For this, the state has to adopt measures to conserve and improve the fish harvesting techniques.

Dr. Sharan D. Crur (2003), in her research paper "One sea Few Fishes – many trawless – A Case Study of the Ramponkars in Goa", describes the traditional fishing and "sustenance" of the fishing industry in Goa. The developmental trajectory has to ensure the survival and livelihood of approximately 30225/40000 odd fishermen who inhabit the 71 fishing villages in the 8 fishing talukas in Goa, and even the context of metamorphosis that is taking place in the sector following the package of economic reforms that were initiated towards the end of the twentieth century.

Dr. I. Satyasundaram (2008) in his research paper "Marine products: In Tapped Potential (Marine products: UN Japed Potential) Status". Though India has over 1.3 million hectares available for shrimp forming, the current exploitation is only 14 percent. Effective exploitation of the deep sea fisheries potential is a challenge, because the heavy capital investments needed for this purpose India are capable of producing value added products by adopting world class technical know-how. The MPEPA acts as a national organization for regulation, development and promotion of machine products exports.

S.S. Guledgudda, M.D. Mathur, M.T. Dodamani (2009) in their research paper "Indian Fisheries Sector: Production Trends and Export Performance" (Fishing Chimes, April 2009, Vol. 29, No. 1) reviewed the growth performance, aspects of potential markets and prospects for the export of Indian fishery products in the framework of the pre- and post reforms W.T.O regime geographical distribution of these exports and current trends in Indian marine products exports.

P.S.B.R. James (2009), in his report "Responsible Fisheries – Key to Conservation Management and Development of Fisheries in India", suggests some strategies, including policy issues for ensuring responsible fisheries practices in the country.

S.K. Naik (2003) "Challenges for Sustainable Capture Fisheries" studied capture fisheries development in India through an organized activity. All efforts to achieve sustained development of fishery have naturally come from the managers of the resources who can ensure "Sustained development growth" by using appropriate "management techniques"

G.P. Ghandhi (2005) examines "Marine Products Exports" from India, i.e., shrimps, which constitute 20 percent of the world's imports in the main stay in India exports of marine products. Frozen shrimps continue to be the largest item of exports, which accounts for 6.9 of total exports of marine products from the country.

Dr. Satya Sundaram (2005) opines "Aquaculture: Fluctuating Fortunes" means to be adopted to increase the competitiveness of prawn culture business in India. Measures are intended to reduce the incidence of infections, as a long gap between crops would prevent horizontal transmission of the disease through water.

Dr. R.K. Jana and Dr. R.K. Sena (2007) presented an exhaustive study on overwhelming growth, which concludes that recent technological advancement have transformed aquaculture to such a height that institutional finance through private and Nationalized sector holds the key to the sources and growth in the entire fisheries sector. Over a period of time assumed credit support through commercial banks and NABARD must be ensured for making aquaculture a commercially viable enterprise.

S. Nagireddy, V. Ramabhandrsraju, M.U.S. Nagireddy (2007) authors have discussed "Fresh Water Fish Framing Practices in Andhra Pradesh". There are many imponderables in the fish farming practices in A.P. which are low technologically oriented and are predominantly manner based.

Albert G.J. and Mathias Halwart (2007) the growing and production of farmed aquatic organisms in caged enclosures has been a relatively recent aquaculture innovation. To date commercial age culture has been mainly restricted to the culture of higher value compound feed fed fish species, including Salamis, most major marine and fresh water carnivores fish species and an ever increasing proportion of omnivorous fresh water fish species, including Chinese carps, tilapia, colossoma and catfish. The cage aquaculture sector has grown very rapidly during the past 20 years, and is presently undergoing rapid changes in response to peruses from globalization and growing demand for aquatic products in both developing and developed countries.

Subramanian, Mohanta and Neelam Kamarpant (2003) Importance of Aquaculture in the sustainable marine fish production have attempted to discuss India's fishery resource production through aquaculture. The advantage of marine culture is increasing the sustainable marine fish production. The potential fish and shell organisms of commercial importance and different types of marine culture methods employed in culturing the species are enumerated. The role, played by the mangroves and its preservation and eco friendly utilization for augmenting marine fisheries, has been focused as well as holistic approach in integrated costal zone development, which would enhance the sustainable development of marine fisheries.

S.K. Naik (2003), in his research paper "Challenges for Sustainable Capture Fisheries", presents an exhaustive study on development of fishery have naturally come from the managers of resource in the state or the administration. As manager of the resource can ensure "sustained development as sustained growth" using – appropriate "management" techniques. Measures like enacting law to meet the biological environmental social and economic aspects of the fishery is the first step towards achieving.

Shamila Monteiro (2009) studied "Farming of Catla and Rohu in Brackish Water Ponds: A Success Story Reviewed the Methods", in order to popularize the culture of Indian major carps. The experiment shows that farming of Catla and Lohu can be taken up successfully in the brackish water farms during monsoons. Shrimp farmers, therefore, can be encouraged to take up farming of IMC (Indian major carps) during monsoons in their ponds, instead of keeping their farms idle. They can also be encouraged to try out fish farming along with scampi for better returns.

Satya Sundaram (1999) "Sea the limit" opines that India has not fully exploited the 200 nautical miles of its exclusive economic zone to earn foreign exchange and increase employment opportunities through export of marine products. The marine product sector remains under tapped since it is an important source of foreign exchange earnings. Fuller utilization of this sector should be given a priority. However care should be taken to minimize the adverse effects on employment of fisher folk and also ecology.

Rama Prasad (2003) "Sea Food needs value addition" states as more and more importing countries demand marine products processed under high hygienic standards, the Indian sea food industry needs to modernize its processing facilities to meet international quality standards.

I. Satya Sundaram (2001) "Marine products – exports hold the key" examined the potential for export of Indias marine products in enormous but it has not been able to exploited due to a variety of constraints. There is an urgent need for modernization of the marine products industry. It will add to the quantum and the quality of each catch. Fishermen can afford to procure the necessary equipment and repay the debt through the export of value added products.

Joe. D'Souza (2002) "Fishing woes to the fore", the fishing scenario is rather murky, there is urgent need for the government to regulate the size of the mesh or else in the years to come one may not find much fish in the sea.

Das and Ashutosh Mishra (2003) "Impact of W.T.O. on Indian Fisheries", in their research paper discussed Indias approach to fisheries is based on the conviction that the emerging world trade regime is both a challenge and opportunity. Liberalization of world trade in fisheries has opened up new vistas of growth. Fisheries is one of the areas in which India has an inherent strength to dominate the global markets.

S.A. Mohite and A.S. Mohite (2008) "Marketing of Fish and Fish Products – Dominant Role of Fishermen", generally fishing is carried out by the men folk, while processing and even marketing of fish a commodity much in demand mostly done by the women folk. The involvement of fishermen starts as soon as the boats reach the fishing harbours landing points.

Women have proved to be the best suited to impart the knowledge about the post harvest technologies such as sorting, icing, packaging and preservation. Further processes such as adequate analysis of consumers tastes, availability and cost of raw material, price of final products, storage facilities, quality control and marketability have to be made available to the women. This is essential to facilitate creation of self/ cooperative employment opportunities to coastal women.

Rajendra K. Majumdar (2006) "Value-added Fish Products" states consumer perception of food has changed through ages. Value-added is the most talked about phrase in the food processing sector. A large number of value-added and diversified fish products both for domestic and export market can be developed.

Surapa Raju (2008) "Fish Retailers – Study in Twin Cities of Hyderabad and Secunderabad", observed the retail fish marketing system in the twin cities of Hyderabad and Secunderabad where there is a lot of demand for fish and where a large number of fish retailers are concentrating. He studies socio-economic conditions or retailers belong to different categories. Study revealed that most of the fish retailers who own shops earn profits that are higher than those secured by the other categories, i.e. cycle/scooter vendors and roadside vendors. Male fish retailers earned more compared to the female fish retailers, etc.

Anjanikumar (2003) "WTO, Food Safety Standards and Regulatory Barriers : Implication for Indian Fisheries Exports", in his research paper examined the implementation of various food safety standards on fisheries exports in India and analyses the costs and benefits of compliance, with these standards and regulations. Maintaining high quality food and fish should be propagated as a strategy to stay ahead of the other competing countries in the world market.

1.15 OBJECTIVES OF THE STUDY

The present investigation has designed specific objectives to pursue the research, and they are as follows:

1. To understand the study area, physio-economics aspects with population and settlement characteristics are the base to the investigation of a theme for the research,
2. To study distribution of fishery resources in the study area from a spatial perception point of view,
3. To observe the different mechanism adopted for exploitation of fishery resources,
4. To study the fish marketing activities in the study area,
5. To identify the problems of fishing industry and fisherman community,
6. To prepare a planning strategy for further development of fishery resources.

1.16 DATA BASE

The present study is based on both primary and secondary data. In order to achieve the objectives of the study, the necessary primary data are to be obtained from different departments and other individual respondent, associated with fishing and marketing activities in Goa.

The required secondary data has been collected from Directorate of Fisheries, Government of Goa, Fishery Survey of India. Marine Products Exporters Development Agency (M.P.E.D.A), Indian Council of Agricultural Research Complex (I.C.A.R.), Old Goa, Exporters Inspection Agency, National Institute of Oceanography (N.I.O.C.), Dona Paula Panaji, Goa, and several other Government departments and Libraries (Central, SCC, NIO, KUD, Goa University) internet browsing, books journals National and International, and reports published by Govt. and non government organizations, individuals, etc.

1.17 HYPOTHESIS

In the present study, the following hypotheses have been formulated and they are as follows:

1. the fisheries Resources are the out come of physical environment rather than socio-economic activities in the region,
2. the marine fisheries production is higher than the inland fisheries,
3. the modern physical infrastructure has encouraged fishing activities rather

than the traditional system,

4. the local market determine the economic status of the fisherman community, the global markets determine the economy of the state Goa

5. the planning strategies for the fishery activities are not only for prosperity of the fisherman community, but also related to the development of the region.

1.18 RESEARCH METHODOLOGY

The Present study has employed the nearest neighboring method for spatial distribution of inland and marine fish landing centers. The fish markets have been also studied by using the same method. The study also encompasses six parameters, and has calculated their percentages, share of each parameter of each taluka. On the basis of percentages, the ranks have been determined. In turn, the total ranks have helped to determine the composite index. On the basis of composite index, six zones have been identified, and accordingly, analyzed with the help of choropleth maps. The taluka-wise and parameters-wise collected data has been compiled, and the tables have been prepared accordingly.

1.19 SELECTION OF THE STUDY AREA

Development of Marketing of Fishery Resources of Goa

Fish assumes scientific significance to the people of Goa, as it forms one of the important items of daily food of more than 90 percent Goa's population. Fishing serves as a means of livelihood to a large number of people of Goa. About 60 percent of the total active fishermen are engaged in marine fishing, while the remaining 40 percent are engaged in inland fishing activities.

In addition to fishing, the number of ancillary and subsidiary activities have grown around fish harvesting, which helps in earning considerable amount of foreign exchange through exports of various fish and fish products and domestic income by supplying fish to neighboring states in the country.

The pace of fishing industry in Goa was becoming somewhat slow, because the state could not get benefits of the first and second five year plan, as it was under the Portuguese rule. However, after the liberation progress has been quite encouraging, and shows rich potential for development of the fishing industry in Goa. In recent years, farmers in the coastal region are turning away from the traditional agriculture, and taking up aquaculture, and therefore, the industry is growing rapidly. The scope for development of fisheries in the state is unlimited because the field is still a virgin one which urged me to undertake study and such a type of resource. As a result the topic "Development of Marketing of Fishery Resources of Goa" has been selected.

1.20 SCOPE OF THE STUDY

The scope of the study is very wide, and is mainly focused on "Development of Marketing of Fishery Resources of Goa", comprising conceptual aspect of Resources in general and fishery resource in particular with regard to History, importance, global, national and state scenario, marketing and functions of marketing, fish marketing and types of marketing.

The study also deals with coastal morphology and the various conditions existing in Marine and Inland fish landing centers and their distribution in Goa. Production and marketing of fishery resource has been the main consideration, i.e., Inland fish production and Marine Fish Production over the years. The local and International trade in fishery products has been analyzed.

Fishery sector provides employment opportunities to more than 5 percent of total working population. Therefore, a primary survey has been conducted on the prospects and problems of fishermen and traders activity in the study area. The role of Directorate of fisheries, Government of Goa, has been examined in promoting and sustaining growth of fishery resources, marketing as well, in order to make fishery sector an important profit making industry in the study area.

1.21 DESIGN OF THE STUDY

The present investigation deals with the geographical perspective of Development of Marketing of Fishery Resources of Goa, and has been arranged systematically into seven chapters in order to understand the entire work carried out. The chapterization is as follows:

The second chapter, which is related to the spatial distribution of fish landing centers, focuses on brief introduction to spatial distribution, coastal geomorphology, geological characteristics, i.e., coastal morphology, fluvial farms, fluvial marine farms, marine farms, Aeolian farms, type of fish landing centers have been analyzed, taking taluka as a unit of analysis. The nearest neighboring techniques are employed to understand the spatial pattern. The temporal distribution of major and minor fish landing centers are also paid the attention.

The fishery resource potentialities of Goa form the base chapter, and hence an attempt has been made in the third chapter to discuss the different aspects related to the fisheries. The introduction, factors influencing development of fishery resources, kinds of fisheries, fish species diversity, methods of fishing related to both marine and inland. The talukawise distribution of fishing gears and temporal trend of fishing gears and crafts have been analyzed. Inland fisheries and marine fisheries have been discussed in detail.

The researcher has resolved to study the marketing activities of fishery resources in the fourth chapter. The fifth chapter has made special effort to study the international trade, along with local trade. The local trade system in different markets are also highlighted.

The fifth chapter deals with planning and strategies with special reference to problems of fishermen community, traders, the role of Government of Goa, private agencies and other leading organizations in sustainable growth of fishery resources and development of fishing industry in Goa as well as trade in fisheries.

The sixth chapter is a concluding chapter confined to a brief summary of all the chapters. Also some specific observations and recommendations for spatial development of fishery resources and its marketing has been discussed.

Chapter-2

SPATIAL DISTRIBUTION OF FISH LANDING CENTERS IN GOA

2.1 INTRODUCTION

Distribution has always been considered as the fundamental step in all the geographical analysis. Through the history of the development of the discipline, the study of distribution has formed its essential core. The identification has been continuously emphasized and reiterated by generation of geographers.

Despite the formulation of a large number of new definitions of geography, the realization of the importance of distribution has never declined whether defined as the science of relationship between man and land or as a study of aerial differentiation or again as spatial organization and spatial interaction, distribution is considered as of fundamental significance. The study of distribution derives its significance from the fact that it is through distribution that the complex, spatially varying area specific relationship between phenomena can be discovered. Similarities and differences among areas which constitute the principle theme of geographical studies are ultimately identified from the distribution of a phenomenon, or a cluster of phenomena. It is one of the foundations of distribution that the first structure of explanation in geography is established (raised).

2.2 SPATIAL PATTERN

The concept of spatial pattern is one of the most important concepts in geography. Search for pattern in the distribution of different phenomena on the surface of the earth has been one of the main pursuits of geographers, planners and diplomats/ policy makers. Cole and King (1968) opine that geography is the science that is mainly concerned with distribution of elements that occur on the surface of the earth and with the variations of distribution through time and space.

The distribution of fish landing centers and their size and nature are closely related to physical, environmental, economic factors and government policies. Geographer's more keen interest in this concept has attempted to find some explanation to the distributional patterns. Reddy (1973) points out that to discern these patterns, to make an objective analysis and to accomplish comparative studies of spatial as well as temporal variations of distribution and their qualification are not only of fundamental importance but also pre-requisite in all studies concerned with space.

An attempt has been made to understand the spatial distribution of fish landing centers, i.e., marine as well as inland with several aspects, i.e., physical, geological, social and economic in the study area.

2.3 COASTAL GEOMORPHOLOGY

Introduction

The coastal zone is the interface between land and water bodies where interaction between natural processes and human activity are most common. The coastal Zone Provides land for settlements, urban development, tourism, agriculture, industry, transportation location for leisure and recreational activities and also has rich fishery potentialities with marine ecology. It is a geologically, chemically and biologically dynamic environment that is subject to considerable natural visibility. It is quite but natural that the coastal zone is dynamic in nature, has paved the way for variations in its functionality to manage the coastal environment. It is the need of the hour to have accurate, corporative and timely scientific data for mainly eco-balance.

The study region is situated along the central-west coast of India. It has a 104 kilometer coastline composed of alternately situated headland estuaries, bays and world famous beaches. It geographically extends from 14° 53° 54° to 15° 48° 00° North Latitude and 73° 40° 33° to 74° 24° 13° East longitude between Western Ghats escarpment and the Arabian Sea.

The coastal zone of Goa offers a host of attractions to exhibit its aesthetic beauty to the tourists from all over India in particular and foreigners in general. It is not only known for the tourist attractions, but also for fishery activities to promote the economy of the region. The natural ports are acting as a place for loading and de-loading either the raw materials or the finished product along with the navigators and visitors. It is need of the hour to protect the coastal environment while ensuring the development is therefore very clear. In order to have a broad perspective of the morphogenetic set up of various coastal landforms, the geomorphologic study of Goa's coast is very essential.

Geological Background

The Pre-Cambrian Dharwar and Cuddapah rocks occupy a large area in Goa. Quartzites, phyllites, chlorite, schists, gneisses, granites, metabasalt, metagrawackes and branded hacmatite quartzite represent the Dharwar rocks, Strata of Cuddapah age occur along the northern coast of Goa from the Terekhol River to the Marmugoa headland. These unfossiliferous metasediments are profusely intruded by basic dykes. The regional trend of Pre-Cambrian rocks in NNW-SSE in the north and veers to East

GOA – COASTAL BELT

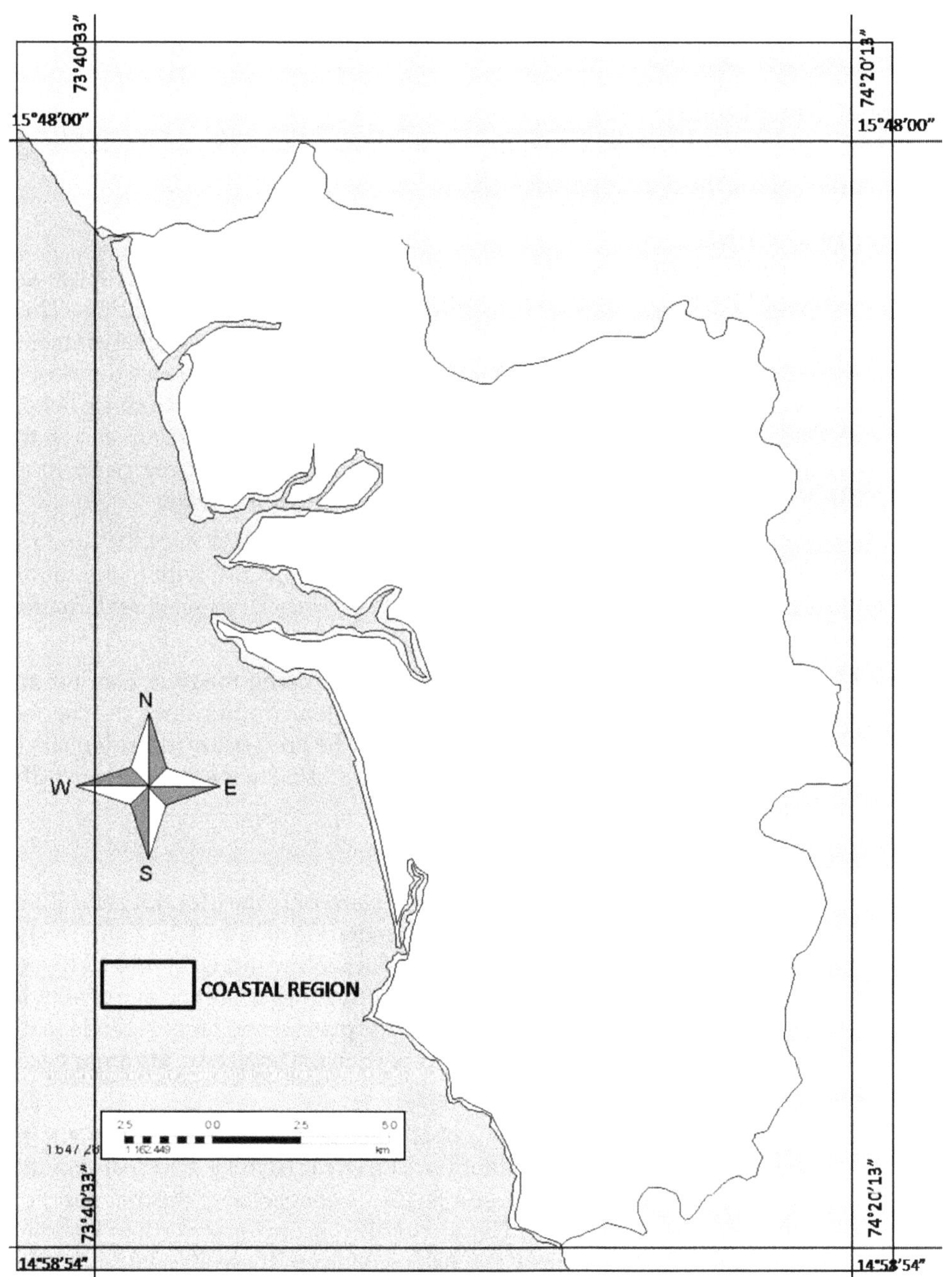

Map-2.1

to West in the south. In a small area in the northeast, the Deccan traps cover the ancient crystalline formations. The Pre-Cambrian rocks are covered by 30-40meters lateritic on the coast that thins to the interior. Laterite beds are also reported in the estuaries of Chapora, Mandovi and Zuari at 20 to 27 and 34 meter below the hydrographic zero.

Coastal Morphology

The following are the important physical characteristics that have been identified and studied along the coastal belt of Goa:

A) Fluvial Forms

i) Drainage : The important rivers draining the area from north to south are Terekhol, Chapora, Mondovi, Zuari, Sal, Talpona and Galgibag. They originated from the Sahyadri ranges and have reached their base level of erosion within a distance of 20-30 kilometer. They are under tidal influence for most of their length towards the coast. The rivers are mostly straight with sharp bends, and meandering is a rare phenomenon. The area is mostly characterized by the dendretic drainage pattern in which the streams branch irregularly in all directions and at almost any angle to form a large trunk stream.

The drainage is predominantly in two directions: one is NW or NNW, and the other, East to West that follows the regional trend and coincides with major faults, fractures and joints. At several locations along the rivers large scale reclamation projects have narrowed the course of the rivers.

ii) Channel Bars : The deposits of sand and gravel along the river channel are designated as channel or meander. Such deposits are found along the channel of Terekhol, Chapora, Sal and Talpona rivers. The presence of meander bars in the convex side of the channel indicate the migration of the river towards the convex end.

B) Fluvial-Marine Forms

i) Estuaries : Out of seven major rivers in the state, only the Mondovi and Zuari rivers have developed large estuaries. The estuarine mouth of the Zuari river is abutted by prominent plateau heights of Bambolim and Cabo in North and Cortalim, Dabolim and Marmugoa in the south with a small enclave of alluvia. The Mondovi estuary mouth is abutted by the plateau heights of Agenda in the North and to the South by the reclaimed portion of Panaji city, Miramar beach and Cabo.

a) Estuary Islands : These islands are probably rocky exposures and are extending peripherally, because of heavy siltation. The Divar Island is a typical example of this type of alluvial island. These islands are connected to the mainland via alluvial and sandy extensions. The island of St. Jacinto just north of Dabolim plateaus demonstrations sandy extensions. Many of these alluvial islands are not obvious features of the landscape, because roads and bridges now link them.

b) River Terraces : These are topographic surfaces marking the former levels of valley floor or flood plain. Terraces reflects variations in base level and changes are some of the important factors bringing in changes in the river channel dynamics. Horizontal or nearly horizontal alluvial terraces on either side of the rivers are about 2 to 5 meter above the present high water level and are covered by alluvial deposit.

c) Tidal Flats : The lower reaches of all the rivers are tidal, sea water encroaches during high tide and submerges the low lying areas adjacent to rivers forming tidal flats. These are extensive, horizontal, marshy or barren tract of land that are alternately covered and uncovered by the rise and fall of the tide. The soil of the tidal flats consists of silty-sand or silty-clay, with abundant organic matter. The old tidal flats cover wide stretches along Terekhol, Chapora, Mondovi and Zuari rivers. The agricultural lowlands, in the coastal tract are old tidal flats formed by sediments deposited along the principal rivers.

C) Marine Forms

i) Mesas : The prominent and impressive landforms along the coast consists of laterite capped mesas often extending 25 to 30 kilometer inland. There is usually an abrupt fall along the coast presenting a cliffy coastline. The average height of mesas varies from 0 to 100 meter above the sea level. They occupy mainly three levels i.e. between 40-60 meters, 60-80 meters and 80-100 meters. The laterite cover ranges in thickness from 10 to 40 meters. Some of the laterites show relief joints and foliations, indicating in-situ weathering of the rocks. The coastal plateaus are generally separated by alluvial or sandy stretches, while in the interior, the lowlands are also lateritic, and at times a plateau of laterite may be followed by another lower level of plateau.

ii) Cliff : Active and abandoned cliffs are most prominent features of marine erosion. Recessions of the active cliffs, which are innumerable along this coast, is ascribed partly to back wearing of rock, masses and to under cutting by waves giving rise in the process to caves, overhangs and the height from 5 to 40 meter. Such cliffs are located along Goa coast at Anjuna, Vagator, Marmugoa, Cape Rama, etc. Equally high abandoned cliffs upto 1 to 2 kilometer inland have been also identified. These are no longer experiencing wave attack as a result of relative drop of sea level. The isolated sea stack show litho logic and structural relation with rocks, a little inland along the coast. At some places, the sea stacks have been worn down and are exposed only at low tides.

iii) Wave Cut Terraces : These are marine abrasion platforms found at the base of the cliffs and headlands. Along the northern sector of the Goa coast the platforms are cut in quartzites and laterite, while in the southern part, these are in metabasalt and gnessic and granite. Along some beaches of the northern sector of Goa, Conglomeratic laterite is exposed forming wave-cut terraces. Some of these terraces are in the inter-tidal zone, but others are above the present high waterline, indirecting earlier higher sea level.

iv) Near-share Islands : The important islands off the Goa coast are Pikene, Grande, Kambarian, St. Jacinto, St. George and Kenko based on lithological and structural identity between northland and the islands. It appears that the near islands represent the detached portions of coastal head lands now isolated from the retrograding coast due to rise in the sea level or due to local tectonic movement.

v) Beaches : This unit is restricted to narrow belt, ranging from a few meters to 100 to 200 meters along the coast. A greater part of the zone forms inter-tidal region. It has been found that most of the beaches are backed by narrow littoral terraces, which are in turn, backed by abandoned cliffs. It is seen that the unprotected areas facing the sea are marked by long stretches of straight sandy beaches with gentle gradient. On the other hand, along rugged shoreline, beaches are mainly limited to the strips at the heads of bay or coves, variably known as bay head beaches, pocket beaches or crescent beaches. Beach sediments consist mainly of quartz along with feldspars and the heavy minerals. They are represented by graths of medium to fine sand class. The backshore sediments are generally coarser.

vi) Beach Rock : Beach rock on the west coast has been described by many geologists. Beach rock has also been observed at several localities along the Goa coast. This rock has almost the same general composition as the loose beach sand. It is familiar to well cemented rock, consisting of mineral grains, rock fragments and calcareous debris, cemented by calcium carbonate. It is formed only in the inter-tidal zone or spray zone, and occurs in thin beds (up to 3 to 4 meters thickness). Along Goa coast, most of the beach rock exposure are found above the high water level about 2 to 4 meter above mean sea level. Their occurrence beyond the present high water level indicates the variations in the sea level.

vii) Beach Ridges : A number of beach ridges are identified far beyond the present shore line. These are a set of very low sandy ridges roughly parallel to the present shoreline, which is indicative of a continuous retreat of the sea. Thus each of them was a shore line in the past. As these ridges are clearly pointing out the regression of the sea, they serve as a major evidence for sea level lowering or regression of sea in these areas.

D) Aeolian Forms

i) Coastal Dunes : Wind is the most important geomorphologic agent with respect to its influence in generating wave and associate loading on the shoreline and in transporting sand inland from the beach and controlling the patterns of sand dunes. A series of sand dunes, which vary in size and degree of stability, occur parallel to the coast. Within these dunes, a set of old and recent or active dunes can be distinguished. Recent dunes are located close to the beaches and rise to a height of 10 to 15 meters. They are generally devoid of vegetation or have only a very thin veneer of vegetation. The older dunes landscape depicts a series of low ridges more or less parallel to the coast separated by swampy depressions. The dunes in the interior are now stabilized by a growth of thick vegetation. These dunes extend to about 1 to 2 kilometer inland and indicate the accretion to the shoreline.

The coast with broad beaches, old and recent dunes abandoned cliffs and old beach ridges indicate that the coast has been progressing seaward and is fairly stable. At the same time, the hilly region of the coast with near shore islands, cliffs, caves, sea stakes, wave cut terraces under formation and the rocky promote rises of the coast is under constant attack of the waves, causing the erosion, and as a consequence, the retrogression of the coast. The presence of lateritic led forming the abrasion plat forms in the inter-tidal zone as well as at a depth of 27th to 35 meter below the present sea level giving rise to the drowned valleys. Other strong evidences of coastal emergence are the presence of alluvial plains, adjacent to river beds as well as surrounding hillocks, which were once islands, present of marine sediments 2 to 7 kilometer inland, abrasion plat forms well above the present sea level and the beach rock along the coast. The present sea level is neither higher nor lower but represents an intermediate level. Such information can be of vital importance to fisherman community and coastal zone planners in effective optimum sustainable exploitation of fishery resources as well as managing the coastal zone.

Coastal Topography : The study area has 104 kilometers of coast, and stretches from Pernem to Canacona. It has been classified into a coastal tract consisting of beaches, sea cliffs, promontories pocket beaches, estuaries, dunes, hard rock wave cut plat form etc. Indian Naval Hydrographic charts 214 and 215 indicate that the continental shelf off Goa is relatively wide with a 50 meters counter depth, occurring 35 kilometers offshore, 100 meters at 80 kilometers and 200 meters at 100 kilometers away from the coast. The sea bed consists of silty clay at water depth less than 50 meters and sandy silt from 50 meters to 100 meters water depth. Beach sediment mainly consists of quartz along with feldspars and other heavy minerals. They are represented by medium to fine sand well to moderately sort.

2.4 SPATIAL DISTRIBUTION OF FISH LANDING CENTRES

Fish landing center plays a vital role not only in the coastal economy but also equally providing necessary facilities to fisherman in exploitation of fishery resources and marketing of fish and fish products in the study area. These are the locations providing the spatial linkages for collection, transportation and marketing of fish and fish products along with coastal belt. Therefore, an attempt has been made to study the spatial distribution of fish and its landing centers.

Table-2.1 : Talukawise Marine Fish Landing Centres in Goa (2007-08)

North Goa	South Goa
Major Centres	**Major Centres**
Malim-Bardez	Kharivado-Vasco
Chapora-Bardez	Cutbona-Salcete
	Betul-Salcete
	Talpona-Canacona

GOA – MARINE FISH LANDING CENTERS

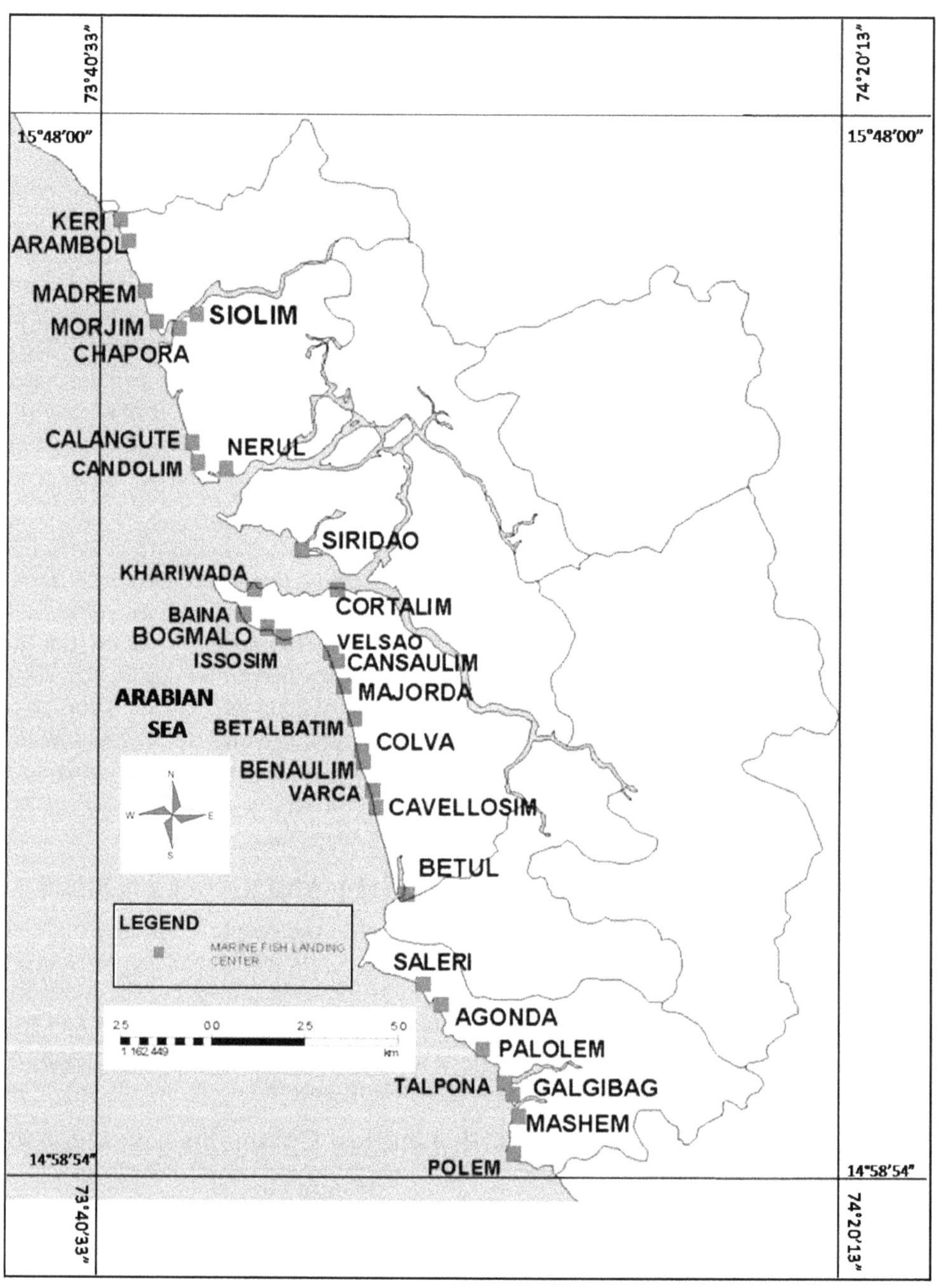

Map-2.2

Medium Centre	Medium Centre
Siridao-Tiswadi	Cansaulim-Marmugoa
Morjim-Pernem	Baina-Marmugoa
Siolim-Bardez	Velsao-Marmugoa
Arambol-Pernem	Benaulim-Salcete
	Colva-Salcete
	Cortalim-Salcete
	Agonda-Canacona
	Palolem-Canacona

North Goa	South Goa
Minor Centres	**Minor Centres**
Nauxi-Tiswadi	Cansaulim-Marmugoa
Mandrem-Pernem	Izossim-Marmugoa
Calangute-Bardez	Majorda-Marmugoa
Candolim-Bardez	Bogmalo-Marmugoa
	Cavelossim-Marmugoa
	Varca-Salcete
	Saleri-Canacona
	Nuvem-Matvem-Canacona
	Galgibag-Canacona
	Pollem-Canacona
	Mashem-Canacona

The following table shows the spatial Distribution of Inland fish landing Centers in Goa.

Table-2.2 : Spatial Distribution of Inland Fish Landing Centers in Goa

TALUKAS	TALUKAS
Bardez Taluka	**Tiswadi Taluka**
Siolim	St. Estevam
Badem	Cumbharjhua
Colvale	Diwar
Camurlim	Chorao
Chapora	Mandur
Aldona	Goa Velha
Pomburpa	Neura
Britona	Agaciam
Nerul	Curcae
Verem	Bambolim
Candolim	Salvo-de-Mundo
Canacona Taluka	**Ponda Taluka**
Colomb	Bandora
Patne	Shiroda
Kiniebaga	Borim
	Durbhat
	Madkai

GOA – INLAND FISH LANDING CENTRES

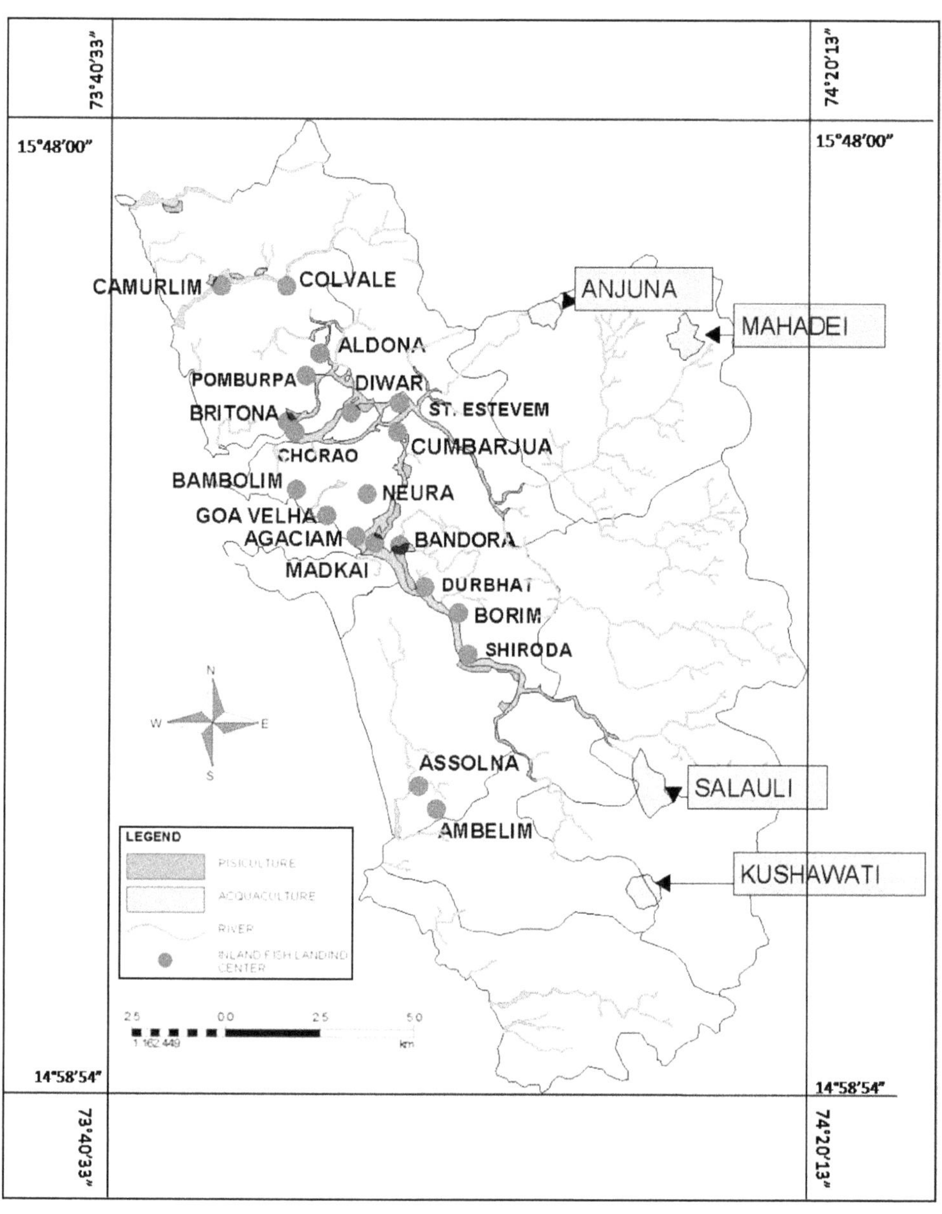

Map-2.3

TALUKAS	TALUKAS
Marmugoa Taluka Chikalim	**Bicholim Taluka** Ammona Piligao
Salcete Taluka Assolna Ambelim	

TALUKA-WISE DISTRIBUTION OF FISH LANDING CENTRES

The study area has been broadly classified into two or is North Goa and South Goa. The analysis has been made as for the administrative zones.

I. Bardez Taluka – North Goa

1. Chapora – Which lies on the coast of Goa, is 23 kilometers from Goa's capital city of Panaji located in the shadow of the Portuguese fort, on the opposite north side of the Vagator. In fact, the location of Chapora Fish landing center is confined to muddy southern share of river Chapora estuary. Dependent upon fishing and Boat building, it has to a great extent retained a life of its own independent of tourism.

   Chapora is basically a fishing village. So a large number of fishermen in their traditional attire can be seen there. There are around 1100 households in Chapora. The total fishermen population is 5500 out of which 259 are actively involved in marine fishing activities. There are altogether 75 mechanized trawlers, involved in fishing operation to enable a large amount of fish catch and economic benefit to fishermen in this area. The following factors are also provided at these centers:

i) Fishing Jetty 30 meter,

ii) Water Supply and illumination (Electricity),

iii) HSD Pump,

iv) Parking Space for Vehicle,

v) Auction shed for Traders and Fisherman.

The Chapora Fisherman Marketing Co-operative Society Ltd., is undertaking the business of fish through Auction shed and pressure the fish in cold storages and thus fish is supplied all over the States too. The space for main jetty has been provided by the state government through fisheries department (Fisheries Department, Govt. of Goa) and Captain of ports.

2. Malim Jetty (Bardez Taluka) North Goa – Is seen across the River Mondovi, located very close to the Panaji, capital city of Goa. There are around 2200 households at Malim. 350 trawlers are operating from this jetty, though many of the trawlers stay away from Malim. Total fishing population is around 8000 at Malim, of which active fisherman are 4500. The following are the facilities provided at Malim jetty complex:

i. 139 meters Fisheries jetties,

ii. Auction Sheds (2 Nos),

iii. Workshops (2 Nos),

iv. Office for Fisheries Society,

v. Canteen Facility,

vi. HSD Pump (2 Nos),

vii. Over head reservoir for water supply,

viii. Net mending shed,

ix. Parking Space for vehicle,

x. Sulabh Toilet,

xi. Illumination Facilities,

xii. Ice plant/Cold storage.

The following are the other fish landing centers concentrated in Bardez Taluka, and their facilities provided by Directorate of Fisheries, Govt. of Goa and Captain of Ports Authority of India.

3. Orda Candolim – Fish landing platform cum ramp. It is a small fishing village. Only 84 fishermen are engaged in fishing. There are 28 registered Canoes at Candolim.

4. Sinquerim – Fish landing Platform cum ramp.

5. Vagator – Fish Landing ramp. This is an important fish landing center of the Inland.

6. Badem Assagao – around 136 fishermen are engaged in fishing.

a) Fish landing platform cum ramp,

b) Net mending Shed,

c) Illumination facility.

7. Gudem Siolim – There are around 286 fishermen involved in marine fishing.

a) Fish landing platform cum ramp,

b) Net mending Shed,

c) Illumination facility.

8. Oxel Siolim – There is 55 registered canoes at Siolim-23(MC) 32(NMC).

a) Fish landing platform cum ramp,

b) Net mending Shed,

c) Illumination facility.

9. Verem Tarir – Fish landing cum ramp. This is an important inland fish landing center.

10. Carmulim – Fish landing platform cum wherein more than 380 fishermen ramp are engaged in fishing.

11. Firguem Bhat, Nerul – Nerul is located in North Goa as a picture village. It shares its boundaries with the historic village of Verem and Reis Magos. The green hills, the blue water of River Mondovi and the cool breeze of the Arabian Sea made Nerul a hot favorite of the Portuguese officials during the colonial days. The Portuguese gently come to the Nerul beach to escape from the heat of the Goan Summers. With the launching of modern tourism, the Nerul beach has become a favorite picnic spot and salt water bath for the locals and tourists.

The population of Nerul is predominantly Catholic. Though agriculture is the basic activity of the village, fishing is an important occupation of the people. There are 95 registered canoes at Nerul, out of which 85 are mechanized canoes. The total fishing population is around 700, and the number of active fishermen is 205. This center is having a fish landing platforms cum ramp.

II. Pernem Taluka – North Goa

Morgim – Morjim Comprises a picturesque portion of Goa 103 square meter long shoreline in the north. The water of the Arabian Sea and the Chapora river provides a rich-breeding zone for a variety of fish. The people are involved in artisan fishing. There are 54 registered canoes, out of which 34 are mechanized and 20 are non-mechanized. Some fisherman from Morgim also own trawlers, but they operate from Malim jetty. Morjim is a fishermen village. 543 active full time fishermen are engaged in marine fishing activities. Pernem Taluka is also having a fish landing platform cum ramp at Keri extreme northern part of Goa. This center has around 70 active fishermen population, and there are 20 registered fishing trawlers, engaged in marine fishing and significantly contributing towards fishery resource production.

III. Tiswadi Taluka – North Goa

Sirdao – Is situated near the Zuari estuary. Srida Beach is a Shell collectors haven with its assortment of oyster and pearl shells. Sirdao is part of Tiswadi taluka, and 350 active full time fishermen are engaged in fishing operation, and 55 canoes are registered, out of which 35 are mechanized and 20 are non-mechanized in this village. The total fishing population in this village is 400. Some fishermen from this area own trawlers, which are kept at Malim jetty at Siridao. The following are the important facilities provided in various parts of the Tiswadi taluka:

1. Avian Caranzalem- Fish landing platform cum ramp,
2. Marvel Donapaula- Fish landing platform cum ramp,
3. Odxel Talegao Fish landing ramp,
4. Acado St. Estevao- Fish landing platform cum ramp,
5. Golwada Kumbharjua- Net Mending Shed.

South Goa

IV. Salcete Taluka – South Goa

1. Betul – Betul is a place located in the Salcete taluka. This place is a one hour journey (distance) from Margo, the commercial capital of Goa. Betul, known for its beautiful beach, is situated along the southern part of Goa, which is

comparatively a small beach. It is located along the southern end of the Sal river in Goa. There are around 1509 active fishermen engaged in marine fishing. This appears to be the biggest fishing population village, because fishing is the primary occupation here. There are around 30 mechanized trawlers engaged in marine fishing operation.

2. Cortalim : is one of the important jetties, and therefore, it is known as Cortalim fisheries complex. There are 30 mechanized trawlers registered for fishing operation, and around 500 fishermen are engaged in marine fishing activities. The following are the facilities provided at Cortalim Fisheries complex:

a) 34 meter jetty with frontage 8 meters,

b) Net mending shed,

c) Water supply and illumination,

d) Auction shed,

e) Parking space for vehicle.

3. Cutbona – Cutbona is one of the important jetties in South Goa, falling into Salcete taluka. There are 270 mechanized trawlers operating through this jetty. Cutbona is a very important fishing village, which has 911 active full time fishermen engaged in marine fishing to earn their living and to ensure greater fish supply in the region and to the nearest major market centers and to minor market centers as well. The following are the important facilities provided at Cutbona fisheries complex:

a) 180 meter Fishing jetty,

b) 144 meter length landing platforms,

c) HSD pump facility,

d) Water supply and illumination,

e) Approach by pass link road,

f) Parking area for vehicle,

g) Net mending shed,

h) Maintenance workshop,

i) Sulabh toilet,

j) Overhead Tank,

k) Kiosk.

4. Assolna – Is a minor fish landing center, which has the following facilities for fishing operation:

a) Fish landing cum ramp,

b) Net mending shed.

5. Colva – Colva village lies nearly 6 kilometer away from Margao. The 25 kilometer of coastline makes Colva an important fishing village. The following ate the facilities provided at Colva:

a) Sub Office fisheries,

b) HSD pump,

c) Fish drying platforms 30 nos.,

d) Illumination and water supply.

The Colva beach starts from Bogmalo in the North to Cabo de Rama in the south. The Colva beach was once the favorite weekend gateway for the upper class elites of Margao. They often headed to Colva for the Mundanca or the change of Air.

Ambelim (Band) – Is one of the important fishing villages of Salcete taluka. It has 71 active fishermen population engaged in marine fishing activities. There are very few number of Canoes registered for fishing operation, mostly of traditional type and therefore, this village has the following facilities provided by- the Directorate of Fisheries Govt. of Goa:

a) Fish landing platform cum ramp,

b) Net mending shed.

6. Canacona Taluka – South Goa

a. Tolpona – Is one of the important fishing areas in Canacona taluka. This taluka has total fishermen (active full time) population around 166, and there are altogether 50 mechanized trawlers and 10 non-mechanized registered Canoes for marine fishing activities to ensure good supply of variety of fish and fish products.

Tolpona fisheries complex has the following facilities:

i. 30 meter jetty,

ii. Fish landing ramp,

iii. Sub office fisheries,

iv. Auction Shed,

v. Illumination and water supply facility.

Apart from Tolpona, other areas have facilities for fishery resource development.

b. Saleri – Fish landing cum ramp. This center has 245 active fishermen engaged in marine fishing, substantially contributing towards development of fishery resources. There are 41 registered Canoes of which 32 are mechanized and are non-mechanized, encouraging fishing operation in this area of Canacona taluka. This center is also having a Net Mending Shed.

c. Nuvem Khola – This is an important fishing village from Canacona taluka wherein 170 full time active fishermen are engaged in marine fishing. There are around 12 registered Canoes, out of which 5 are mechanized and 7 are non-mechanized or traditional. This center is also having a Net mending shed provided by Directorate of fisheries, Govt. of Goa.

d. Agonda – Agonda is emerging as one of the important marine fishing centers in Canacona taluka wherein 240 active fishermen are involved in marine fishing activities. There are 22 registered Canoes, out of which 16 are mechanized and

6 are non-mechanized, helping in fishing operation. This center is having a Net Mending shed provided by the Govt.

e. Kiniebag – Kiniebag is an important inland fish landing center wherein 250(245) active fishermen are engaged in fresh water fishing activities. There are registered centers for inland fishing in Canacona taluka. This center has a Net Mending Shed to facilitate fishermen.

f. Galzibag – Is a fishing village of Canacona taluka. It has around 140 active fishermen population, and there are 8 registered Canoes out which 3 are mechanized and 5 are non-mechanized, contributing towards fishery resources development.

g. Pollem – is an important fishing village wherein 316 active fishermen are engaged in marine fishing, and there are 11 registered Canoes, out of which 9 are mechanized and 2 are non-mechanized, encouraging fishing operation in this area.

h. Mashem – Mashem is a small fishing village. The fishermen population is around 36 who are actively engaged in marine fishing activities. There are no registered mechanized Canoes operating in this area. Only a few traditional crafts are involved in fishing activities. There are 5 registered non-mechanized Canoes existing at this center. Mashem is yet to be developed as an important fish landing center, provided that the local youths participate on large scale and the Directorate of fisheries takes appropriate measures in years to come.

i. Pallolem – is an important fishing village from Canacona taluka. This center is located in extreme southern part of Goa which is the second last fish landing center of the Goa state where fishery resources is concerned. There are 785 active full time fishermen engaged in marine fishing, and there are 70 registered Canoes, out of which 55 are mechanized and 15 are non-mechanized or traditional type, immensely contributing towards fishery resource development in Canacona taluka. Pallolem is the largest fishery resource center known for quantity and variety of fish collection in this area.

j. Pollem – Pollem is a small fishing village situated at southern tip of Goa in Canacona taluka. The fishermen population is about 316, who are actively involved in marine fishing activities, substantially contributing towards fisheries production and its supply in the market. There are 11 registered Canoes, out of which 9 are mechanized and 2 are non-mechanized, helping in fishing operation. Thus Canacona taluka is playing a vital role in production of fishery resources and their supply marketing as well in the State of Goa.

7. Marmugoa Taluka – South Goa

a. Kharivado – Kharivado falls in Vasco-da-Gamma, which is a Key shipping center. Vasco-da-Gamma has quite an extensive fishing fleet and boat building/ Shipyard located in and around the harbour. Kharivado, which is known in local language as Kharevaddo, is a fishing area. There are around 235 mechanized trawlers and 40 motorized Canoes. Total fishing population of Kharivado is 2798, actively involved in marine fishing, and therefore, this collection helps to supply variety of fish to various market centers in Goa.

b. Velso – is a village in Marmugoa taluka. There are 1319 registered Canoes, out of which 79 are mechanized and 60 are traditional crafts, engaged largely in Marine fishing to ensure good catch and supply. The total fisherman population is around 489 at this center, actively involved in fishing activities throughout the year except during ban period of monsoon.

c. Baina – is an important fishing village from Marmugoa taluka. It is the 3rd largest center for marine fishing activities. The active full time fishermen population is around 1620 in the whole of Goa. There are 115 registered canoes, out of which 63 are mechanized and 52 are non-mechanized, stimulating fishing operation in Marmugoa taluka. A variety of Sea fish has been caught from this center and sold in the major and minor market centers of Goa.

d. Cansaulim – is a small fishing village with a population of 47 who are active full time fishermen engaged in marine fishing. There are 85 registered Canoes out of which 45 are mechanized and 40 are non-mechanized. This center immensely contributes towards the fishery resource development and its supply to varied market centers in Goa and outside of Goa.

e. Bogmalo – is an important fishing village along the coast of Goa with total fishermen population around 288, actively involved in marine fishing activities. There are 59 registered Canoes, out of which 24 are mechanized and 35 are traditional type encouraged in fishing activities on large scale in Marmugoa taluka.

f. Benaulim – is the second largest center for marine fishing in the whole of Goa as there are around 1707 active fishermen, engaged in fishing operation throughout the year, except ban period during monsoon. This Center is known for the richest collection of fish and its supply to various parts of Goa. There are 46 registered Canoes, out of which 12 are mechanized and 34 are non-mechanized, stimulating fisheries production. Thus overall Marmugoa makes significant contribution towards fishery resource development of Goa.

8. Ponda Taluka – South Goa

Goas inland water bodies have tremendous potentiality for development of fisheries resource. Among various taluks Ponda taluk has emerged as an important leading producer of fresh water fish and as an inland fishing taluk in the study area.

There are five inland fish landing centers located in different parts of Ponda taluk and they are equipped with various facilities which are as follows;

a) Durbhat – There are 350 fishermen engaged in inland fishing. Durbhat fish landing center has been equipped with one fish landing platform cum ramp and a net mending shed.

b) Madkai (Caranzod) – Fishermen 370

i) Fish landing platform cum ramp,

ii) Net mending shed.

c) Madkai (Tonk)

i) Fish landing platform cum ramp,

ii) Net mending shed.

d) Bandora Kariwadda – There are 80 fishermen engaged in inland fishing activity, and this area is characterized by one fish landing ramp cum platform.

e) Kundai (Manaswada) – It has one fish landing ramp cum platform. About 70 fishermen are engaged in inland fishing operation in this region.

3.5 NEAREST NEIGHBOUR ANALYSIS

In the regional studies, the study of spatial as well as temporal variations in distributional pattern of fish landing centers, settlements and market is of great importance. According to Watson (1955) geography it self is a discipline in distance, and according to Cole and King (1968) Geography is the science that is mainly concerned with the distribution of element that occur on the earth surface and with the variations of distributions through time and space.

Nearest Neighbour Method

In order to get a statistical measure of the pattern of distribution of fish landing centers, the nearest neighbour method of analysis has been used. Nearest neighbour analysis method was first developed by plant ecologist Clark and Evance (1954) who were concerned chiefly with explaining distribution patterns of various species of plants and trees on the surface of the earth. The geographers applied this technique to people, factories, settlements, markets and other items to explain the location pattern. The analysis is based on the ratio between observed and actual mean distance that might occur under random conditions. The principle followed in this method is straight-line measurement of distance separating point from its nearest neighbours in the space.

L.J. King (1969) opined that study of distribution pattern and location pattern is essential to geographers. They should elaborate "pattern of geographical aspects, which are spaced in a particular region or area". Geographers rely upon the optional identification of patterns. Recently Mulimani (2006) has employed this method for spatial distribution of market centres in Raichur district and Hugar (1994) has applied the same technique for analyzing the distributional pattern of markets in Gulbarga district of Karnataka state.

When Rn=1, the pattern is completely random, high Rn scores represents various degrees of dispersion, with Rn=2 resulting from the limiting regular pattern based on a triangular lattice. Generally, the Rn value is greater than one indicative distribution leading towards a uniform and Rn value less than one indicate leading towards clustered. This nearest neighbour analysis was modified and revised by N.B.K. Reddy. According to him the higher Rn scale is 2.15 and it represents maximum dispersal and lowest Rn scale is to represent a critical distribution in between these two extremes. This critical stage is conspicuous by its absence in the Rn scale given by clear in Evans and others. Regular dispersion and exigencies or geographic conditions.

Example: Strategic considering the guidelines suggested by N.B.K. Reddy for his revised Rn scale. The author has further modified the scale as follows;

Table- 2.3 : Nearest Neighbor Index

Rn Score	Distribution Pattern
2.149	Uniform
1.25	Leading towards uniform
1	Random
0	Cluster

Spatial distribution of marine as well as inland fish landing centres in Goa has been observed with the help of nearest neighbouring technique in the following ways.

Marine Fish Landing Centres in Goa

$$Rn \quad \frac{\bar{Do}}{\bar{De}}$$

$$\bar{Do} \quad \frac{\text{Total Measured Distance}}{\text{Total No.of Settlements}}$$

$$\frac{18}{31}$$

$$0.58$$

$$\bar{De} \quad 0.5\sqrt{A/N}$$

Where as

A Area

N No.of Settlements

$$0.5\sqrt{\frac{3702}{31}}$$

$$0.5 \quad 10.92$$

$$5.46$$

$$Rn \quad \frac{\bar{Do}}{\bar{De}}$$

$$\frac{0.58}{5.46}$$

$$0.10$$

$$Rn \quad 0.10$$

The Map-2.4 displays the results based on nearest neighbour analysis of marine fish landing centres in the study area. As per the spatial distribution is concerned the calculated Rn value is 0.10, thus the marine fish landing centres are distributed in the linear pattern in the study region.

GOA – MARINE FISH LANDING CENTRES

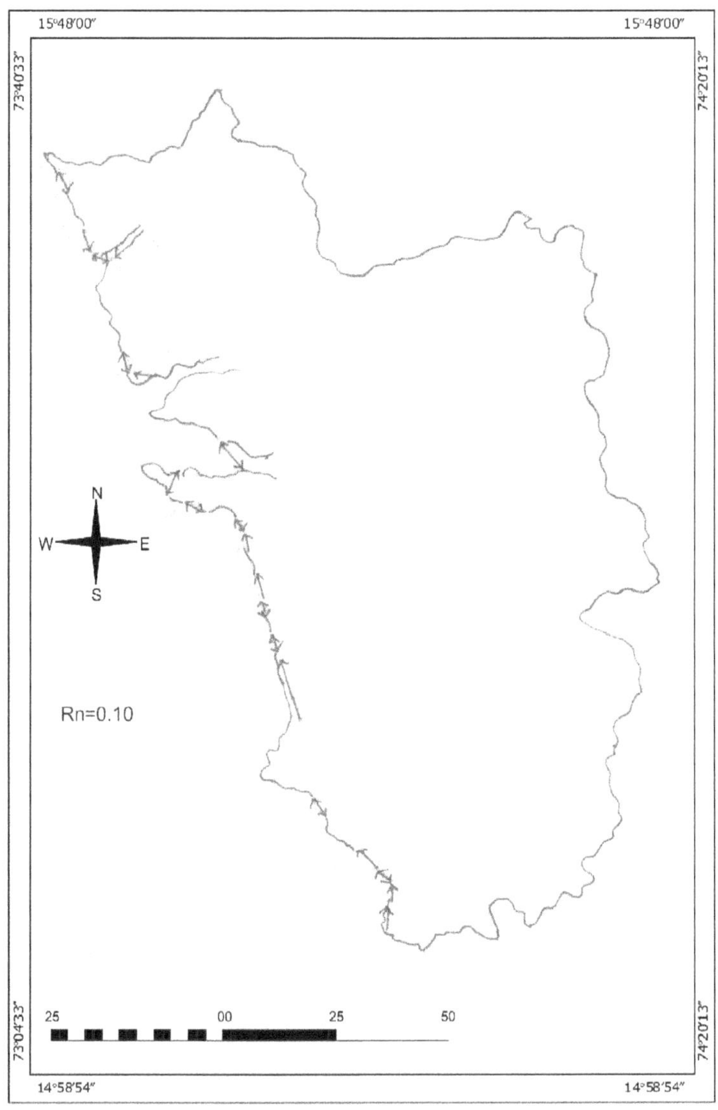

Scale 1cm = 1.25 km Area = 3702 sq.km.

Map-2.4

Inland Fish Landing Centres in Goa

$$Rn \quad \frac{\overline{Do}}{\overline{De}}$$

$$\overline{Do} \quad \frac{\text{Total Measured Distance}}{\text{No. of Settlements}}$$

$$\frac{13.12}{20}$$

$$0.65$$

$$\overline{De} \quad 0.5\sqrt{A/N}$$

Whereas

A Area

N No. of Settlements

$$0.5\sqrt{\frac{3702}{20}}$$

$$0.5 \quad 136$$

$$6.80$$

$$Rn \quad \frac{\overline{Do}}{\overline{De}}$$

$$\frac{0.65}{6.80}$$

$$0.09$$

$$Rn \quad 0.09$$

The Map-2.5 displays the results based on nearest neighbor analysis of inland fish landing centres in the study area. As per the spatial distribution is concerned the calculate Rn value is 0.09, thus the inland fish landing centres are distributed in the form of clustered distribution pattern in the study region

GOA – INLAND FISH LANDING CENTRES

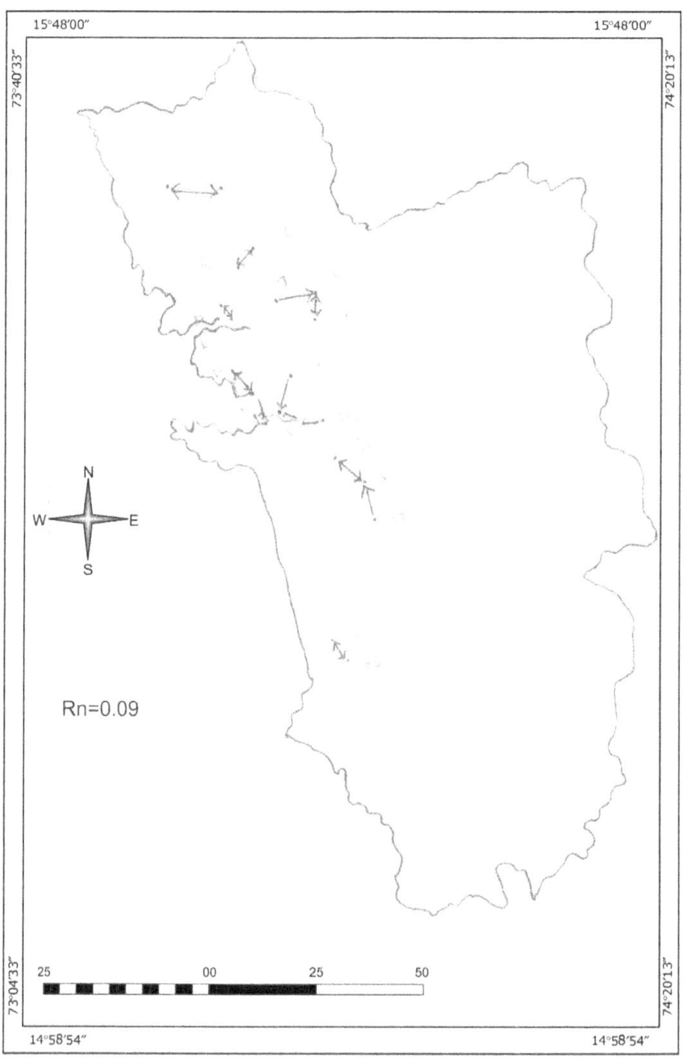

Scale 1cm = 1.25 km Area = 3702 sq.km.

Map-2.5

2.6 TEMPORAL DISTRIBUTION AND DEVELOPMENT OF FISH LANDING CENTRES

Introduction

It has been observed that in the study area, fish landing centers have been unevenly distributed along the coast as well as inland water bodies. These centers are gradually developed under the supervision of Directorate of Fisheries, Government of Goa over the years. The present study is focusing on three different periods of time with the interval of a decade, i.e., 1980-90, 1991-2000 and 2001-2010. This decadal variation is the component of temporal aspects. Hence, the temporal distribution of fish landing centers have been analyzed.

The Directorate of Fisheries, Government of Goa has taken the initiative to develop marine fish landing centers as well as inland in different parts of Goa by way of construction of jetties, extension of salty platforms, construction of ramps, auction sheds, net mending sheds and so on.

The table 2.4 indicates that the trend of temporal distribution and development of fish landing centers in Goa during the three different periods during 1980-90, 1991-2000 and 2001-2010.

There are 41 fish landing centers, which have been developed and are functioning in the seven taluks of the study region. There are only three centers, i.e., Jetties have been developed in only two taluks, namely, Bardez has seven and Salcete one. In the same decade, there 29 ramps have been constricted. Bardez taluk has the highest number of geographers. They are not only concentrating on the spatial aspects, but are also focusing on temporal variations. The temporal variation is the indication of the changes or developments that have taken place over the decades.

Hence, an attempt is also made in the present study to analyze the fish landing centers from temporal point of view.

Table-2.4 : Temporal Distribution and Development of Fish Landing Centrs

Taluk	1980-90		1991-2000		2001-2009		Total	%
	Jetty	Ramp As NMS	Jetty	Ramp As NMS	Jetty	Ramp As NMS		
Pernem	-	-	-	1	-	-	1	2.4
Bardez	1	-	7	11	-	-	19	46.3
Tiswadi	-	-	-	5	-	-	5	12.1
Mormugao	-	-	-	-	-	-	-	-
Salcete	1	-	1	4	-	1	7	17
Canacona	1	-	-	3	-	-	4	9.7
Quepem	-	-	-	-	-	-	-	-
Ponda	-	-	-	5	-	-	5	12.1
Total	**3**	**-**	**8**	**29**	**-**	**1**	**41**	**100**

Source: Direactorate of Fisheries, Goa (2010).

Bardez taluk has 7 jetties and 11 ramps, followed by Tiswadi and Ponda five each. The Salcete taluk has four, followed by Canacona three, Pernem taluk has only one ramp cum fishing platform. During this decade 1991-2000 there were 8 jetties construction and 29 ramps established, and they are functioning effectively. In the year 2001-2010 hardly any ramps were developed, except only one at Salcete taluk.

As compared to three decades, the 1990 decade has flourished as far as jetties and ramps are concerned. In the beginning of 1980s, ramps are rarely found. The very specific reason is that the Department of Fisheries had taken effective steps to ensure fishery cultivation to prosper in nature naturally without much human interference. Therefore it was good in the beginning of this decade to promote and develop fish landing centers in various taluks.

Pernem taluk accounts for 2.4 percent of the development which is very low. There was only one ramp cum platform, which was constructed at Kerim in 1990-91. Further no developments have taken place in this region.

It has been observed that maximum developments have occurred in Bardez taluk, which represents 46.3 percent of fish landing centers, ramps construction activities. In 1982, fish landing platform was developed at Chapora. Fisheries jetty at Malim was established in the year 1991-92. Malim jetty was constructed with necessary facilities, i.e., auction shed, cold storage, etc. Several ramps have come up during the period 1991-2000. One ramp each developed at Sinquerim, Firguem Bhat Nerul, Orda Candolim, Nagator, Ram and net mending shed at Badem Assagoa, Gudem Siolim and Oxel Siolim and ramps have come up in Camurlim Tarir as well as at Verem Tarir. In all, there are around 11 ramps related construction activities that have taken place during the 1991-2000 period. Since then there are no further developments in this region, as it appears to be saturated with existing developments. Therefore, not a single construction activity related to fish landing centers has come up during the period 2009-10. Large scale developments reflected greater amount of fishery resource development.

Tiswadi taluk accounts for 12.1 percent of total construction activities and ranks 3rd in respect of ramps related construction activities. 5 ramps related have taken place during the period 1990-2000. One ramp each is developed at Aivon Caranzalem, Marvel Donapaula, Odxel Taleigao. One Ramp and Net mending shed at Golwada – Kunharjua, and one ramp at Akado, St. Estivan, are the important inland fish landing centers in this region.

Salcete taluk ranks second in respect of fish landing centers development in the study area. It represents 17.1 percent of total construction activities. During the 1989-90, a major fishing jetty was constructed at Cutbona, and it has become an important center of a variety of fish collection, apart from other jetties in the study area. One jetty related and four ramp related constructions have come up during the period 1990-2000, i.e., ramp at Colleawado Assolana, Net mending shed at Golleadando Wado Assolana, Net mending shed, auction shed at Cortalim and RCC pavement to fisheries jetty Cortalim. During 2000-2001, one ramp and net mending shed has been constructed at Ambelim Band which is being used for fishing operation.

Marmugao taluk records Nil development of fish landing centers by Directorate of Fisheries, Government of Goa. Khariwaddo jetty has been developed by MPT Government of India.

Canacona has been the southern most taluk of Goa which accounts for low development of fish landing centeres, i.e. (97%) of total and ranks fourth in fishery infrastructure concerned. In the year 1987-88 one fishing jetty was constructed at Talpona wherein river Talpona meets the Arabian Sea in the west and it is one of the oldest fishing jetty in the study area. One ramp at Saleri canal was developed in 1993-94, Net mending sheds at Agonda and Nuvem Khola were developed during the period 1996-97. Since then no further construction of jetties, ramps and auction sheds has taken place during 2001-2010.

Qucpem taluk represents zero development with regard to fishing jetties and ramps concerned. As it is having the rocky coastal belt wherein sea land interface is very negligible. Hence, no fishing ramp or jetty point has been developed, and scope for subsistence fishing is limited as well.

Ponda taluk is well-known for inland fishing activities in the study area, and ranks third position in the development of ramps concerned. Five ramp constructions took place during the period 1990-2000, which accounts for 12.1 percent of total ramp constructions. One ramp each at Khariwadda Bandora, Kundaim, ramp and net mending shed at Madkai Tonk, Karanzol Madkai and Durbhat. Ponda have been developed to extract inland fishery resources in this region. The decade of 2001-2009-10 records Nil developments of fish landing centers. Still there is scope for construction of platforms, ramps, etc. in order to give encouragement to fishermen involved in inland fishing activities.

Hence, the hypothesis that the fisheries resources are the outcome of physical environment rather than socio-economic activities in the region, has been tested with literature and confirmed the hypothesis.

Chapter – 3
FISHERY RESOURCE POTENTIALITY OF GOA

3.1 INTRODUCTION

Fishing is one of the extractive occupations of mankind older than agriculture. Fish assumes greater significance to the people of Goa as it forms one of the most important items of the food of more than 90% percent of population. In fact, fish forms an integral part of Goan life. Fish and Rice together form a significant staple food. Indeed, people are so fond of fish that no meal is complete without fish dish. The word renowned "Xit Coddi" is almost invariably a part of the daily meal in every Goan household. Subramanian (2000) says, "Per capita fish consumption in Goa is 7.4 kg as against the national average of 5 kg, and the recommended average of 11 kg.

Fishing also serve as a means of livelihood to a large number of people of Goa. Altogether more than 5 percent of total working population are engaged in fishing activities. Apart from tourism, mining and manufacturing fishing industry forms the second largest industry both in terms of employment and income. In addition to fishing, the number of ancillary and subsidiary activities have grown around fish harvesting, which, in turn, stimulates the growth of Goa's economy.

3.2 FACTORS INFLUENCING FISHING ACTIVITIES IN GOA

The following are the various factors influencing the development of fishery activities in different parts of Goa, i.e., Favourable climate. Good demand, Transport, Market centres, Broken coastline, Infrastructure, Inland Fresh water bodies and availability of labour, etc. Their details are as follows:

a) Favorable Climate

Goa experiences tropical oceanic climatic conditions with orographic influence and varied seasons, that is, S:W monsoon season N:E monsoon season, Winter season and Summer season. Generally, the climate of Goa is moist and balanced through out the year. During the Rainy season, the conditions are usually unfavourable when the sea is rough and not suitable for fish catch, but it is most suitable for the growth of fisheries. From September to March conditions are very ideal for fishing activities, benefiting a large number of fishermen community of Goa.

b) Good Demand

There is a greater demand for fish not only in the local areas but also outside the state. 90% of total population consume fish on daily basis. The Density of population is high in Bardez, Tiswadi, Marmagao and Salcete taluka. Therefore, there is greater demand for fish throughout the year, and a large portion of the total catch is marketed in these talukas compared to Canacona, Pernem, Sanguem, Bicholim and Sangquelim. During the lean season, fish is imported from neighbouring states, i.e., Karnataka, Maharashtra and Kerala to meet the needs of consumers.

c) Transport

Transportation is a key factor for marketing the fishery resources. Since it is a perishable commodity, it has to be marketed immediately. Without the development of transportation, the sale transaction will not take place. The state is well connected by National Highways NH-17. NH-4A State Highways, a number of village roads, taluka roads and district roads. Flexible network of transport enables quick and safe disposal of fish and fish products in various talukas. The density of vehicle population, that is, four wheeler and two wheeler has been increasing every year to meet the needs of ever-growing rural and urban population.

d) Market Centres

Goa has well established market centres in both rural as well as urban areas. Mapusa, Panaji, Margao, Vasco and Ponda are the largest market centres of Bardez, Tiswadi, Marmagao and Salecte talukas. These talukas have well established fish marketing centres, where as other areas have minor market centres. Major Towns and cities have largest fish collection brought from different fish landing centres and jetties. Thus fish is traded on large scale regularly, which ensures greater economic stability of fishermen in general and a trader in particular.

e) Broken Coastline

Goa has a 104 kms broken coastline with numerous bays and headlands, which serve as a base for fishing operations. There are many major and minor fish landing centres such as Betim (malim) Chapora, Khariwado, Cutbona and Betal, etc., from where large quantity and variety of fish is caught and made available for distribution and marketing.

f) Infrastructure Facilities

Infrastructure facilities are well established in the study area. The transportation system is well developed and the market centers are well organized cold storage facilities as well as IRS Digital Visual systems are developed at important locations, which helps in increasing catch of fishermen and allows easy channel of distribution of fish and fish products.

g) Investment

The state is well endowed with a large number of financial institutions. Major Towns and Cities of Goa have a large number of Nationalized Banks, Private Banks, Co-operative Banks and Credit Societies, etc. On an average, each village is having at least one financial institution. The people involved in the fishing activities and operating such activities are to be provided the loan facilities from the said financial institutions and invest for this activity. Government is also encouraging by providing financial assistance in the form of schemes, granting the loans with subsidies.

h) Labour Supply

Fish catching and Trading are Labour intensive activities. Therefore, the large supply of labour is required. The state has a number of developing towns and cities. Being an important international tourism destination, there is a large influx of people from neighboring states and other parts of India to these areas to avail employment and business opportunities. Fishery sector of Goa employs largely outside labour rather than local labour. As a result, the density of population of Goa has been increasing year by year. Thus the migration of people is solving labour problem in the study area not only in fishery sector but also in other areas.

i) Fresh Water bodies

There are a number of lakes and fresh water bodies located in different parts of Goa, from where a large variety of fish is caught and sold in different market centers. Besides, there are several short and long seasonal tanks in the villages, used for raising fresh water fish and shrimps.

The noted fresh water bodies are Anjunem reservoir, Selaulim reservoir, Maya lake, Carambolim and Churchodem, etc., facilitating the development of Fresh water fish culture.

3.3 KINDS / TYPES OF FISHERY RESOURCE

Goa consists of mainly Two Kinds of fisheries.

A) Inland Fisheries

B) Marine Fisheries

A) Inland Fisheries

Inland Fisheries of Goa are one of the richest source spread over 250 kms. Two hindered and Fifty kilometers in length. Inland fisheries are divided into two types, that is, Brackish and Fresh water fisheries.

Brackish water fisheries include extensive estuaries or river mouth, a large number of lagoons, back waters and brackish water lakes, etc.

Fresh water fisheries are constituted by great river system, fresh water lakes, vast networking irrigation canals, tanks, reservoirs and ponds, etc.

B) Marine Fishery Resources

Marine Fishery Resources comprising of coastline of maximum 104 kilometers. It is a broken coastline characterized by numerous bays and head lands. Groups of oceanic Islands with numerous creeks, mangroves, swamps and coral reefs are believed to be extensive.

3.4 METHODS OF INLAND FISHING

Inland Fishing is mainly carried out with the help of mechanized and Non-mechanized or traditional crafts in various parts of Goa. The methods of Inland fishing are as follows:

A) Fishing by cast-nets, gill nets, etc.

B) Fishing by stake nets

River fishing is carried out in the creeks of the four major rivers, i.e., Chapora, Moadovi, Zuari sal and Talpona. Creek fishing may be done either by caste nets, river gill nets and barrier nets, etc.

Fishing is also practiced with the help of state (stake) nets in the river creeks, which forms an important type of inland fishing activity. It is pursued throughout the year, except the time of high tides during the monsoons. This type of fishing mainly depends upon the ebb and flow of tides and current of water, which differs from day to day.

The following operations are generally carried on for about eight to ten days in a fortnight. The important characteristics of some of the Inland fishing methods are as follows:

a) Small hand-drawn nets/towel

This method involves the use of the small rectangular piece of mosquito net or a porous cotton towel. If the material is not too long, one end of it may be tied around the neck of collector, and the other side can be used to scoop out fishes while the collector wades through shallow water. Whenever there is a partner, it is better for each to hold one end of the net towel and drag it along in a synchronized manner. The net/towel should be held in such a way that the length is stretched between the neck and hands of between the two collectors. The net is then vertically held, the lower end touching the floor. The depth of water should not exceed the width of the net/towel. When dragged from the deep towards the shallow end of a pond, many fishes are trapped. By a quick turning of the sides to enclose the captured fish, a catch may be successfully made.

This method is very useful in small ponds, streams and river edges if they are not deeper than the knee. It is useful in narrow channels and rock pools where most other methods are difficult. Also, this would be the most suitable method for catching and transferring small fish alive. However, using this device only the small fishes are caught. Larger ones are able to leap over the sides when cornered.

b) Dip nets

Nets on circular or square frames can be used in a number of situations. These nets are fitted with handles of varying lengths. The user may rest on a bank and the net to collect small fish is cast in narrow channels and pools where it is not possible to get into the water. Long handled dip nets are also used in catching small fishes alive. Dip nets are most useful for catching fishes that swim along the surface. They can be used very easily during the night.

c) Drag nets

These are nets of varying sizes, sometimes designed as baskets with a considerable depth. When the corners are tied to ropes and the net is dragged along the bottom of aquatic habitats, many fishes are tapped. Depending on the size and nature of the habitat, a large fish may also be caught without any damage. The difficulty arises only when there is a very swift flow of water and the floor is not even. It is not possible to use this net where there are many submerged rocks.

d) Cast nets

These are widely used devices. Cast nets can catch large sized fishes even in deep and open waters, including where there is rapid flow. The only difficulty is that the fish gets entangled and suffocates soon and many die in the net as they are removed.

e) Gill nets

These are also popular amongst fishermen and researchers. The nets are allowed to drift across lakes and rivers for a period of time. The top is kept close to the surface with floats and the bottom is held down with some weights. Fishes that attempt to swim through the nets are trapped. The disadvantages of the method are that the fishes are mostly dead by the time they are collected, bottom-dwelling fishes are rarely caught, and there is a further danger of the nets drifted away and being lost.

f) Traditional methods

Traditionally, man has used a variety of traps and plant poisons to capture fishes. Each village usually has its own unique traditional device for fishing. While these can also be useful, most of them are aimed at catching as many fish as possible.

g) Hook and line

This is particularly useful in deep and muddy waters. However, it is very time consuming. The advantages are that by using the right bait, certain species may be selectively caught. The disadvantage is that the hook can damage the fish very badly when they are being removed.

3.5 METHODS OF MARINE FISHING

In order to fish, a fisherman generally requires a craft-boat and gear-fishing nets. The gear is either passive or active, depending on the nature of its use. It is passive, if it is used in a manner whereby fisherman waits for the fish to get entangled in the net. When fish are chased, disturbed or by circled, the gear used is called active. John Kurien (1982) examined technological change in fishing and its impact on fishermen in Kerala. He found that there was a considerable increase in production of fish in Kerala due to adoption of mechanization and some traditional matter.

The following are the various techniques adopted for marine fisheries widely developed along coasts as well as in deep seas:

(1) Operation at Beach seines,

(2) Gill net fishing in off shore waters,

(3) Cast net fishing in shallow waters,

(4) Seasonal hook and line fishing,

(5) Trawling,

(6) Purseining .

Their details are as follows:

(1) Operations Beach Seines

A beach seine type of fishing is locally known here as RAMPON Fishing is the main activity. Nearly 42% of the total population are active fishermen engaged in marine fishing undertaking. This type of fishing are solely dependent on it. It is carried on by an organization of twenty five to thirty fishermen, or sometimes even more depending on the size of the net.

The activity is carried out at all the coastal points where the conditions are most suitable such as relative evenness of the sea-bottom, relative freedom from heavy surf, etc. This activity is concentrated in the four coastal blocks at Bardez, Ilhas, Marmagoa and Salcette as the conditions near the coast of these blocks are most favourable for the operation of beach-seines.

(2) Gill Net Fishing

Is another type of marine fishing carried out in deep waters about seven to eight miles away from the shore. The nets have a larger mesh size and are manufactured at home by fishermen. According to their requirements, these nets are mainly used for catching large size fishes, viz., shark, seer fish, pomfrets, skipjack, etc.

Fishing by gills net is relatively a specialized type of activity, and is regularly pursued throughout the fishing season. Since this type of fishing has to be done in deep waters, all the navigational rules and regulations are to be strictly observed by the fisherman.

(3) Cast Net Fishing in Shallow Waters

Perhaps the most primitive term of gear, are used all over India and have the same basic design everywhere. Cast net fishing is done individually in shallow waters with a boat and a net. Sometime, fisherman form a group of two to three persons and row their boat in the midst of the shoal either of mackerel or sardines and catch the fish by throwing the net. The catch is meager compared to the efforts put in by the fishermen. This practice is adopted by the fishermen residing a little away from the sea-coast.

(4) Fishing by Hook and Line

Is carried on only during the peak fishing season. Selected varieties of fish such as seer fish, silver bar, skip jacks, serranus, char menus, etc., are caught with this type of fishing. This involves highly skilled technique and is mainly pursued on the Marmugao coast.

(5) Trawling

The Trawler net is a large bell shaped net, wide at the mouth and tapering to through the body to the closed end, called the cod end. It is an active gear. The mechanized boat which is used to pull a trawl net is called a trawler, and this method of fishing is called trawling. There are many kinds of trawling operations possible. One method of trawling is when the net skims either just below the water surface or in the mid-water. Another method which is more commonly used, bottom-dwelling species of fish and crustaceans like prawns. When carrying out bottom trawling, it becomes necessary to attach with an iron chain called a 'tickler' to the bottom part of the mouth of the trawler net. This digs out the prawns and other bottom dwelling fish from their habitats.

(6) Purseining

Purseining is called an active fishing method that uses a huge encircling net with a small mesh size capable of catching tonnes of shoal fish that is mackerels and sardines and even a tiny baby fish. Purseining is a labour saving device suited for the western countries where labour is scarce and expensive. It catches almost all fish, thereby leaving the traditional fisherman with empty nets.

3.6 TALUKAWISE FISHING GEARS

The following table shows fishing gears registered taluka wise as on 2007-08 by the Directorate of Fisheries Government of Goa.

High - Above 1200 – Tiswadi

Medium - 601-1200 – Bardez, Marmugao, Salcette

Low - 500-600 – Pernem, Ponda, Canacona

From the above table it appears that in all 5542 of gears (different types) have been registered as on 2005 from various parts of the study area at the Directorate Fisheries, Government of Goa by Fishermen to enable their fishing activities to earn their livelihood.

Tiswadi taluka accounts for 23.14% of total gears registered. There are 1281 fishing nets held by the fishermen. Out of which 543 are sea gill nets, 281 trawl nets, 71 river gill nets and 81 are barrier nets respectively. There are also stake nets about 174 and 43 drag nets are used for fishing operations. Thus this region ranks first in the use of all types of fishing gears as compared to other areas in Goa.

Marmugao ranks second in the use of fishing gears. There are 1161 nets held by the fisherman accounts for 20.9% of total fishing gears in the state. There are 621 Sea gill nets, 315 trawl nets, 35 barrier nets, 6 stake nets and 59 Purse seine nets. River gill nets are very low in number, that is, 13 on account of limited scope for Inland fishing activities.

Bardez taluka has 1026 fishing gears, which accounts for 18.51% of total fishing gears in the state. 439 Sea gill nets, 227 trawl nets, 59 Purse-seine nets and 57 drag nets are used for marine fishing activities. 56 river gill nets are used for inland fishing 41 besides river caste nets. Hence, this area ranks third in use of fishing gears for necessary operation.

Table-3.1 : Taluka wise Fishing Gears registered

Sl. No.	Type of Gear	Tiswadi	Bardez	Ponda	Pernem	Canacona	Mormagao	Salcete	Total
1.	Sea gill net	543	439	19	179	404	621	220	2335
2.	Sea Cast net	-	6	-	3	35	-	-	44
3.	River gill	71	56	39	9	9	13	14	202
4.	River Cast	5	41	2	7	13	-	7	66
5.	Rampon net	13	6	-	9	5	14	46	93
6.	Drag net	43	57	-	2	61	12	-	173
7.	Purse Seine	49	59	3	-	5	59	143	315
8.	Trawl net	281	227	-	9	31	315	387	1250
9.	Barrier net	81	51	55	14	30	35	39	293
10.	Sluia gate	21	21	15	-	-	4	12	73
11.	Stake gate	174	57	84	8	-	6	142	471
12.	Kadsari	-	-	6	-	-	-	-	6
13.	Other net	-	6	-	-	-	12	3	18
14.	Orurol net	-	-	-	-	-	19	7	19
15.	Single net .	-	-	-	-	-	51	-	51
	Total	**1281**	**1026**	**223**	**245**	**593**	**1161**	**1013**	**5542**
	%	**23.14**	**18.51**	**4.02**	**4.42**	**10.7**	**20.9**	**18.27**	**100%**

Source : Directorate of Fisheries, Govt. of Goa (2008-09).

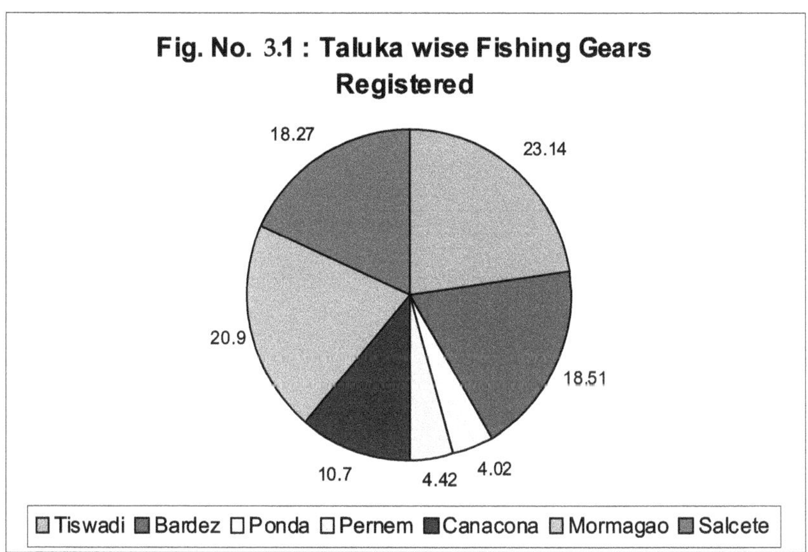

Salcete taluka ranks fourth in using fishing gears, which accounts for 18.27 percent of total fishing gears in the state. There are 1013 fishing nets out of which 220 are Sea gill nets, 387 trawl nets, 46 rampon nets, 143 Purse-seine nets and 39 barrier nets. There are a few River gill nets, that is, 14 and 7 river caste nets are used. There is not much scope for inland fishing activities. Therefore, the use of river gill net and caste is very limited in this taluka.

Pernem taluka records 4.42% of total fishing gears in the state. There are 223 gears, out of which 179 are sea gill nets, 3 sea cast nets, 9 river gill nets, 9 rompon nets and two (2) are drag nets. Fishermen are also using 14 barrier nets to carry out their necessary fishing operation.

Canacona accounts for 10.7% of total fishing gears registered in the state. There are 593 fishing gears, out of which 404 are sea gill nets, 35 sea cast nets, 61 drag nets, 31 are barrier nets, 31 are trawl nets and 9 river gill nets.

Ponda Taluka is known largely for Inland fishing activities. There are 223 fishing gears, out of which 19 sea are gill nets, 39 river gill nets, 55 barrier nets and 84 stake nets.

There are no trawler nets which may be due to limited fishing activities in this area. Therefore, It accounts for 4.02 percent of total fishing gears registered in the state, which is the lowest compared to other areas. It has been observed that the Pernem, Ponda, Canacona taluka have less number of fishing gears on account of limited growth of fishing activities, small size of fishermen population, and lack of basic infrastructure facilities in the respective areas.

Tiswadi, Bardez, Marmugao and Salcete talukas have large number of fishing gears due to a large scale development of fishing activities, active participation of a large number of fisherman population and the presence of basic infrastructure facilities, etc. Thus, the fishermen community is availing itself of all the possible schemes from the Directorate of Fisheries, Govt. of Goa for acquiring all the necessary equipments to develop their fishing activities, not only at the subsistence level but also commercially.

The following table displays the temporal trend of fishing crafts and gears registered in the study area during 1980-81 and 2009-10.

Table-3.2 : Temporal Trend of Crafts, Fishing Gears Registered

Sl. No.	Fishing Crafts	1961 -62	1981 -82	1991 -92	2001 -02	2003 -04	2009 -10
I	Fishing Crafts						
A	Mechanized Boats						
i	Gill Netters (O.B.M.)	-	251	900	460	968	-
ii	Trawlers	-	282	-	255	1134	1157
iii	Liners	4	36	837	-	-	-
iv	Others	-	21	-	601	-	728
	Total	4	590	1757	1324	2102	1885
B	**Non-Mechanized Boats**	**4125**	**2205**	**1900**	**1866**	**1700**	**859**

Source: Directorate of Fisheries, Government of Goa (2010).

The above table shows the trend of fishing crafts and gears registered at Directorate of Fisheries, Government of Goa to carry out fishing activities on commercial scale by the fishermen community in the study area since liberation, that is, from 1961-62 to 2009-10. There has been increase in mechanized trawlers, canoes and non-mechanized canoes till 2003-04, but after that there was sharp decline in the use of non-mechanized canoes, and the use of trawlers stagnated.

Table-3.3 : Fishing Gears Used

Sl. No.	Fishing Gears	1961 -62	1981 -82	1991 -92	2001 -02	2003 -04	2009 -10
1	Purseine	-	-	-	301	355	383
i	Drag nets	-	544	196	171	200	175
ii	Gill nets	-	4688	1925	2269	2450	2625
iii	Trawl nets	-	580	598	1221	1500	1350
iv	Cast nets	-	707	-	55	60	44
v	Trap	-	206	-	286	300	-
vi	Shore seines	-	424	-	93	-	-
vii	Spawn collecting nets	-	142	-	-	-	-
viii	Others	-	832	1342	842	695	1334
	Total	**-**	**8123**	**4065**	**5238**	**5560**	**5911**

Source: Directorate of Fisheries, Govt. of Goa (2010).

The stopping of further registration of mechanized trawlers, the Directorate of Fisheries, Government of Goa to prevent excessive mechanization, has lead to over exploitation of fisheries, destruction of mangroves and pollution of water (sea pollution caused by oil spillages, rejected materials, waste, etc.), which have caused ecological imbalance in the marine eco system of the study area. Aggravating the situation, i.e., depletion of nursery fish fauna, certain species are becoming extinct from Goa's waters over the years.

3.7 FISH SPECIES DIVERSITY IN GOA'S WATER

The study area has a favorable coastline and equable – equitable climate throughout the year. The growth of mangroves along the costs as well as brackish water bodies is influencing the growth of fish fauna on an unprecedented scale. As a result, the state has rich fish species diversity.

Hamilton made a description of 269 species of fishes from the Ganges and its tributaries in Pioneer work "Fisheries of Ganges" Suret (600 BC) made ecological classification of fishes. There are altogether 120 species of fish existing in saline and fresh water bodies of Goa, and they are subsequently trapped and brought at various fish landing centers by Fishermen. Finally, a large variety of fish finds its way into the local markets all over Goa and outside the state and country as well through various modes of transport, i.e., Road, Rail and Water.

Out of the total 120 species of fish, 101 species are marine as well as inland – freshwater fish, while the remaining 19 species are shrimps of fresh water and brackish water.

The important ones are Shark, which is called variously in local language; Mori, Catfish/Sangot Herings, Sardines/Tarlo, Pedwe, Bombay duck/Bombil, Wolf Herings/Kurli, Black silum pampret/Halva, Paplet, Mackerels/Bangda, Tuna/ Wiswan, etc. and remaining 19 species are shrimps i.e. Jinga Shrimp, Giant Tiger Prawan and Prawns of various sizes – Big, Small and Medium. Prominent among the shark fish varieties are sardines and mackerel, which constitute 46 percent of the marine species.

In the given study area, out of one hundred and twenty species of fish, 40 species have been selected for the study of their characteristics, as they are largely used by people of Goa for various purposes. The following are the scientific characteristics of various species of fish found in Goas water.

CHARACTERISTICS OF SELECTED SPECIES OF FISH

Family: Lethrinidae

Scientific name: Lethrinus nebulosus

Common name: Spangled emperor

Local name: Tamso

Field identification characteristics:

Body depth is greater than head length, inner surface of pectoral fin base scaled.

Colour olive green above, paler below, usually 2 or 3 blue streaks radiating from eye scale on the back with a white to blue centre, usually several yellow longitudinal stripes on the sides.

Family: Lutjanidae

Scientific name: Lutjanus argentimaculatus

Common name: Mangrove red snapper

Local name: Tamso

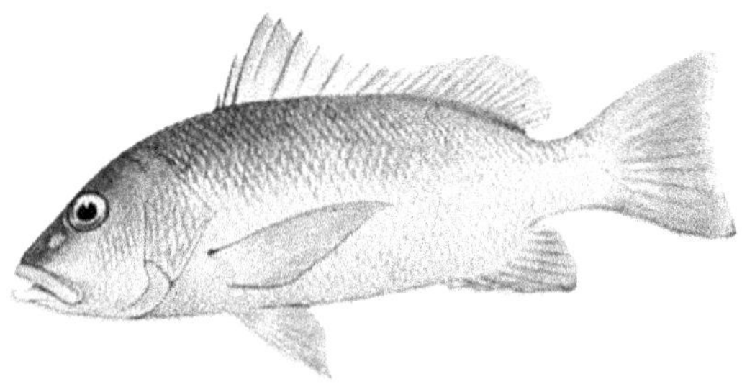

Field identification characteristics:

Longitudinal rows of scales above lateral line parallel to dorsal profile anterior appearing to rise obliquely under soft part of dorsal fin or under posterior part of the dorsal fin.

Scale rows below lateral line horizontal.

Colour red brown, somewhat paler on belly, often a silvery patch in the centre of each.

Scientific name: Lutjanus malabaricus

Common name: Malabar blood snapper

Local name: Palu

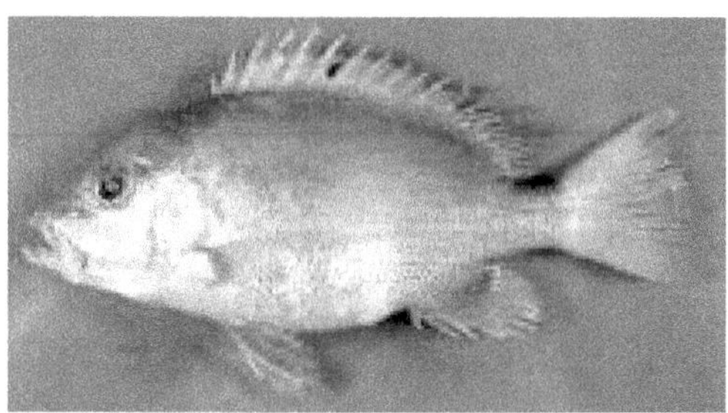

Field identification characteristics:

Head profile straight or concave, dorsal fin with 11 spines and 14 soft rays, anal fin with 3 spines and 8 or 9 soft rays.

Longitudinal rows of scales above lateral line appear to rise obliquely to dorsal profile, those below lateral line horizontal.

Colour deep red in adults, juvenile's red/brown above, with dark longitudinal stripes on body.

Scientific name: Lutjanus johni

Common name: john's snapper

Local name: Palu

Field identification characteristics:

Longitudinal scale rows above lateral parallel to it, and those below lateral line horizontal.

Body silvery green or bronze/red, with a distinct spot on each scale forming a lengthwise series of dark streaks.

A large black present above lateral line at junction of spinouts and soft part of dorsal fin and this is often surrounded by a silvery ring in juveniles.

Scientific name: Thunnus albacares

Common name: Yellowfin tuna

Local name: Bugde

Field identification characteristics:

Two dorsal fins separated only by a narrow interspace, the second is followed by 8 anal fins are followed by 7 to 10 finlets.

Large specimens have very long second and anal fins.

Pectoral fins moderately long, usually reaching beyond second dorsal fin origin but its base.

Back metallic dark blue changing through yellow to silver on belly, belly frequent about 20 broken, nearly vertical lines. All fins and finlets bright yellow, from narrow black border.

Scientific name: Thunnus obesus

Common name: Bigeye tuna

Local name: Rogtado

Field identification characteristics:

Pectoral fins moderately long in large specimens, but very long in small specimens.

Back metallic dark blue, lower sides and belly whitish, a lateral indecent blue along sides, 1st dorsal fin deep yellow, 2nd dorsal and anal fins light yellow, fin yellow edged with black.

Scientific name: Thunnus tonggol

Common name: Longtail tuna

Local name: Rogtado

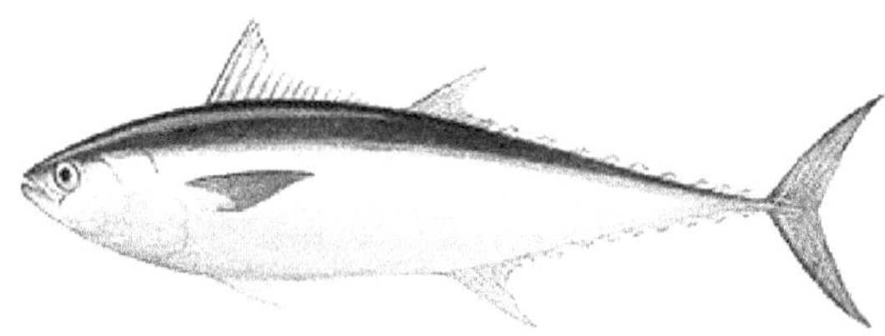

Field identification characteristics:

Two dorsal fins separated only by a narrow interspace, the 2nd is higher than the 1st and is followed by 9 finlets, pectoral fins short to moderately long.

Back blue or black, lower sides and belly silvery white with colourless elongate oval spots arranged in horizontally oriented rows, dorsal pectoral and pelvic fins blackish, tip of 2nd dorsal and anal fins yellow, anal fin silvery, dorsal and anal finlets yellow with grayish margins, caudal fin blackish, with streaks of yellowish green.

Scientific name: Rastrelliger kanagurta

Common name: Indian mackerel

Local name: Bangdo

Field identification characteristics:

Body moderately deep, gillrakers very long, visible when mouth is opened.

Back is blue green, flanks are silver with golden tint, 2 rows of small, dark spots on sides of dorsal fin bases, narrow dark longitudinal bands on upper part of the body and black spots on body near lower margin of pectoral fin. Rests if fins are yellowish to dusky.

Family: Istiophoridae

Scientific name: Istiophorus platpterus

Common name: Sailfish

Local name: Tarso Masso

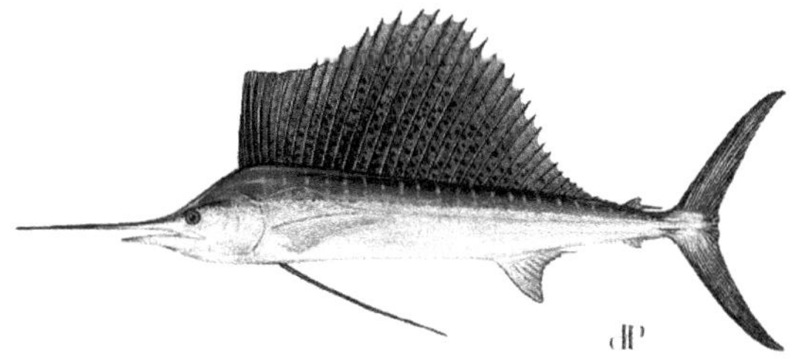

Field identification characteristics:

Body elongate, much compressed. Upper jaw prolonged into a rather spear.

Two dorsal fins, 1st dorsal fin is large and sail-like, pelvic fins very long almost anus.

Body dark blue dorsally, silvery white ventrally, 1st dorsal fin membrane blue, many small black spots, other fins brown-black, more or less 20 horizontally consisting of several pale blue spots on body.

Scientific name: Makaira indica

Common name: Black marlin

Local name; Tarso Masso

Field identification characteristics:

Body elongate, not strongly compressed. Upper jaw prolonged into a rather slender shape very steep.

Two dorsal fins, the 1st long and low posteriorly, the 2nd small, height of anal of the first dorsal fin smaller than body depth.

Pectoral fins falcate, rigid, not foldable back against sides of body, pelvic fins short pectorals, consisting of 1 spine and 2 soft rays. Lateral line single.

Body dark blue dorsally, silvery white ventrally, first dorsal fin membrane blue-black usually unspotted, other fins brown-black.

Scientific name: Scomberomorus commerson

Common name: Narrow-barred Spanish mackerel

Local name: Wiswan

Filed identification characteristics:

Body elongate, rather compressed, lateral line abruptly bent downward below end of second dorsal fin.

Back iridescent blue/grey, side's silvery with bluish reflections, marked with numerous thin wavy vertical bands, juveniles are frequently spotted.

Scientific name: Scomberomorus guttatus

Common name: Indo-pacific king mackerel

Local name: Wiswan

Field identification characteristics;

Lateral line, with many fine branches anteriorly almost straight to below middle of 2nd dorsal fin, and gently bent downward to middle of caudal peduncle.

Colour blue on back, silvery on sides, about 3 irregular rows of dark round spots along sides of body, spinous dorsal fin dark up to 8th spine, white posteriorly with the distal margin black.

Family name: Sciaendae

Scientific name: Dendrophysa russelli

Common name: Goatee croaker

Local name: Dodiaro

Field identification characteristics:

A single barbel on chin, median mental pore at the base of the solid, pointed metal.

Back grey, shading to white on belly, a dark brown band on nape, operele with a blotch, upper edge of spiny part of dorsal fin dark.

Scientific name: Johnius belangerii

Common name: Belangers croaker

Local name: Dodiaro

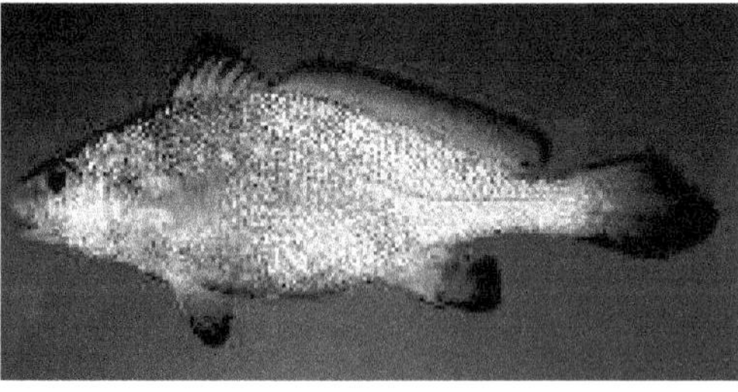

Field identification characteristics:

Rostral pores 5, teeth differentiated into large and small in upper lower jaw teeth villiform.

Body darkly pigment sometimes irregular and concentrated into small bars along back or on dorsal fin, spinous part of dorsal fin black, a dark blot through gill cover.

Scientific name: Nibea maculate

Common name: Blotched croaker

Local name: Dodiaro

Field identification characteristics:

Rostral pores 3, marginal pores 5, notching the edge to produce 3 lobes, mental pores 5.

A distinctive Colour pattern of 5 dark bars extending obliquely from the back to the lower part of flanks, and sixth dark blotch on top of caudal peduncle.

First bar broadest, from nape obliquely backward, lower part of bars narrower and often discontinuous.

Scientific name: Protonibea diacanthus

Common name: Spotted croaker

Local name: Dodiaro

Field identification characteristics:

A big, nearly horizontal and terminal mouth, teeth differentiated into large and small into both jaws.

5 dark bars along back and many small black spots on the top head, upper half of body and caudal fin, pelvic, anal and lower part of caudal fins black. In large fishes the large bars and the small spots are absent.

Scientific name: Hilsa toil/tenualosa toli

Common name: Toli shad

Local name: Pede

Field identification characteristics:

Body fusiform, moderately deep and compressed, belly with a distinct shade keel of silver.

Upper jaw with distinct median notch.

Colour blue/green, flanks silvery, at most a diffuse dark blotch behind gill opening between spots on flanks.

Scientific name: Sardinella longiceps

Common name: Indian oil-Sardinella

Local name: Tarlo

Field identification characteristics:

Belly rounded, with a low keel of scutes.

Colour back blue/green, flanks silvery. A black spot on hind edge of gill cover and a patch on the body behind it.

Family: Psettodidae

Scientific name: Psettodes erumei

Common name: Indian spiny turbot

Local name: Lep

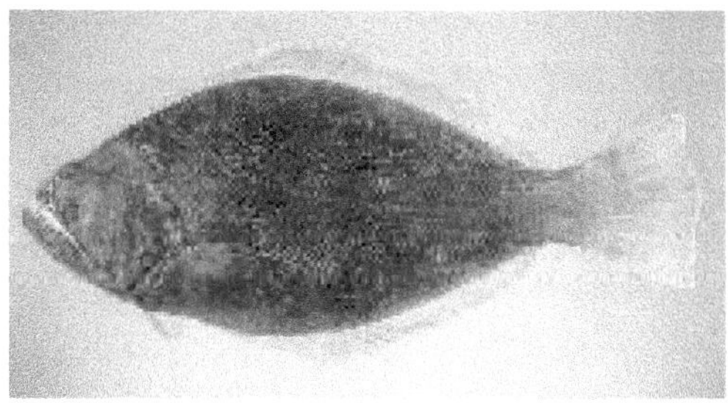

Field identification characteristics:

Mouth large with strong teeth. Both the eyes on the same side, upper eye lying immediately below dorsal edge.

Dorsal fin origin well posterior to eyes.

Body usually brown/grey, sometimes with 4 broad, dark crossbars. Dorsal anal and caudal fin tips black.

Family: Cynoglossidea

Scientific name: Cynoglossus macrostomus

Common name: Malabar tonguesole

Local name: Lep

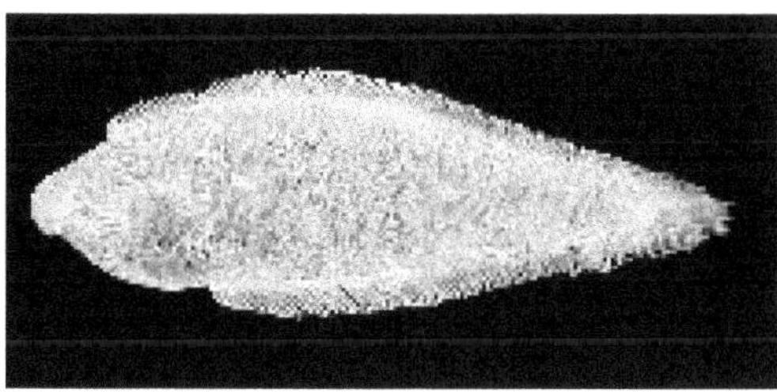

Field identification characteristics:

Eyes on left side of the body, with no space between them.

Two lateral lines on eyed sides but none on blind side.

Colour light brown on eye side with dark brown mottling forming diffuse, irregular cross bands.

Family: Chirocentridae

Scientific name: Chirocentrus dorab

Common name: Dorab wolf-herrings

Local name: Karli

Field identification characteristics:

2 fanglike canines pointing forward in upper jaw, a series of canine teeth in lower jaw.

Upper part of the dorsal fin black, inner face of pectoral fin black at base, some anterior part of the anal fin.

Scientific name: Chirocentrus nudus

Common name: White wolf-herrings

Local name: Karli

Field identification characteristics:

2 fanglike canines pointing forward in upper jaws, a series of canine teeth in lower jaw.

Dorsal fin clear, inner face of pectoral fin black at base, no black on anterior part of anal fin.

Family: Clupeidae

Scientific name: Anodontostoma Chacunda

Common name: Chacunda gizzard-shad

Local name: Gibber

Field identification characteristics:

Body very deep and compressed, belly with a keel of scutes.

Mouth inferior, snout rounded and projecting.

Back blue/green, flanks bright silver, a large jet black spot behind gill opening.

Family: Sphyraenidae

Scientific name: Sphyraena jello

Common name: Pickhandle barracuda

Local name: Tonki

Field identification characteristics:

No gillrakers on 1st arch, upper and lower gill arch platelets rough, but without distinctive spines, scales small.

Colour blue/black or brown above, side's silvery, with a dark pattern of serpentine reaching a little below lateral line, but no inky spots on hind part of body below lateral line.

Scientific name: Sphyraena obtusata

Common name: Obtuse barracuda

Local name: Tonki/Tensori

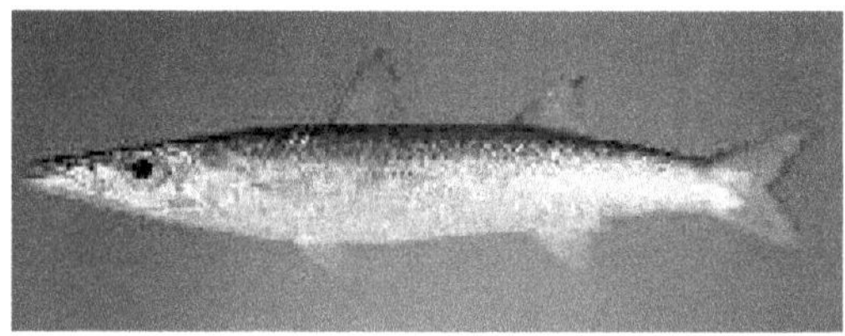

Field identification characteristics:

2 gillrakers on 1st arch, upper and lower gill arch platelets rough, but without distinct spines.

Colour grey/brown with greenish tinge above, sides silvery white without dark bars chevrons, 2nd dorsal anal and caudal fins yellowish.

Family: Carangidae

Scientific name: Parastromateus Niger

Common name: Black pomfret

Local name: Surgunteo/Halva

Field identification characteristics:

Dorsal fin with 4 to 5 short spines (embedded and not apparent in adult).

Pelvic fins absent.

Straight part of lateral line with 8 to 10 weak scutes, forming a slight keel on caudal peduncle.

Family: Stromatidae

Scientific name: Pampus argenteus

Common name: Silver pomfret

Local name: Paplet

Field identification characteristics:

Mouth small, curved downward, maxilla immobile covered with skin and united to cheek.

5 to 10 flat blade-like spines preceding the median fins, dorsal and anal fin with posteriorly elevated lobes, no pelvic fins, and caudal fin forked.

Body silvery white on sides, head slightly brownish, edges of fins are blackish.

Family: Ariidae

Scientific name: Arius caelatus

Common name: Engraved catfish

Local name: Sangot

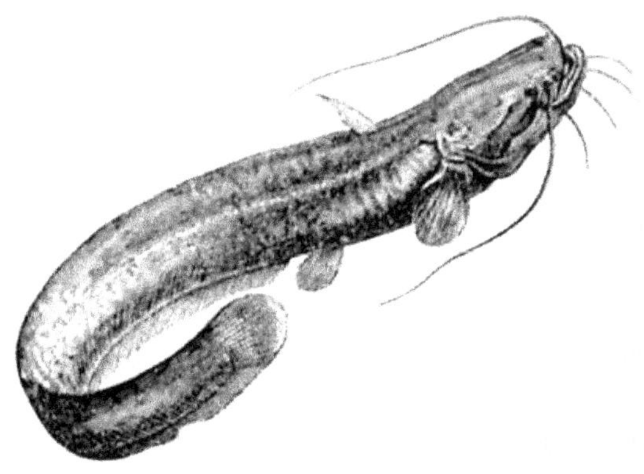

Field identification characteristics:

Head shield rugose and granulated posterior to orbit, especially on the supraoccipital tapering to a wide V towards occipital region, and with a preorbital conical protuberance.

Palate teeth villiform and densely packed in a small, roughly triangular patch on each side.

Tip of the dorsal fin produced occasionally into a black filament.

Scientific name: Arius tenuispinis

Common name: Thinspine sea catfish

Local name: Gaggor-Sangot

Field identification characteristics:

Median longitudinal groove long, narrow, deep, running onto supraoccipital process.

Palatine teeth villiform along outer margin, globular along inner margin, in a single ellipse shoe-shaped large patch on each side, placed far back in the buckle cavity.

Scientific name: Arius thalassinus

Common name: Giant catfish

Local name: Sangale Sangot

Field identification characteristics:

A prominent preorbital conical protuberance tapering as a wide V posteriorly.

Palatine teeth villiform, in 3 patches on each side, forming a triangle, posterior patch longest, patches usually fused (slightly separated in juveniles).

Scientific name: Osteogeneiosus militaris

Common name: Soldier catfish

Local name: Sangale Sangot

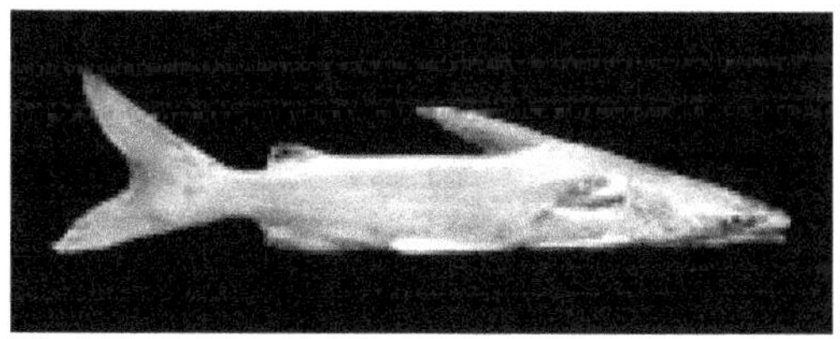

Field identification characteristics:

Only one pair of stiff, semi osseous maxillary barbells absent.

Head shield smooth without granulation or rugose striations.

Top of head and back intense dark blue.

Family: Seyllaridae

Scientific name: Thenus orientalis

Common name: Flathead locust lobster

Local name: Sheword

Field identification characteristics:

Carapace flat, widest in front, distinctly narrowing posteriorly, lateral margins straight, only 2 teeth, one at end, the other in the anterior fourth, posterior ¼ without teeth.

Anterior tooth forming part of the orbit, which is situated at the anterolateral angle of carapace. Fifth segment with a sharp spiniform tooth in the middle of posterior margin.

Colour pale yellowish brown with the granules of a darker brown. Tips of the teeth white. Tail fan with a yellow tinge.

Family: Portunidae

Scientific name: Charybdis cruciata

Common name; Christian crabs

Local name: Kurli

Field identification characteristics:

Carapace broad and flat, 5 to 9 teeth on anterolateral margin, no spine on dactyls.

Distal 2 segments of last pair of legs mare flattened than in anterior legs, dactyl of last pair of legs usually oval.

Body purplish with a large yellow cross, chelipeds purple, spotted with yellow dots.

Scientific name: Portunus Pelagicus

Common name: Blue swimming crab

Local name: Kurli

Field identification characteristics:

1 spine near posterior margin and 2 spines near anterior margin of chelate legs.

Transverse deep brown with numerous creamish spots allover the carapace, chelate and swimming legs, walking legs bluish to dusky.

Family: Veneridae

Scientific name: Meretrix

Common name: Venus shell

Local name: Dhave Khubey

Field identification characteristics:

Large, triangularly ovate, thick with polished and glossy surface.

Colour straw yellow with dark brown band on left edge of shell.

Scientific name: Catelysia opima

Common name: Venus shell

Local name: Khubey

Field identification characteristics:

Small, tick, solid and flattened shell with glossy surface.

Colour pale yellowish brown or straw colored, mottled and purplish grey markings.

Scientific name: Papiha textile

Common name: Venus shell

Local name: Tisrio

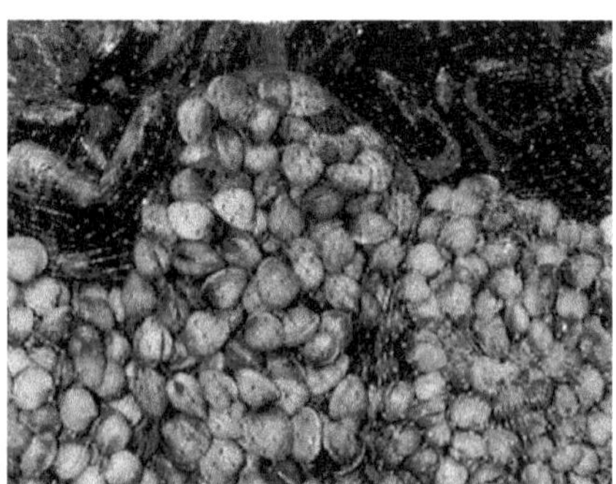

Field identification characteristics:

Shell greatly elongated, thick and heavy with smooth and glossy surface.

Colour pale yellowish white marked with purplish grey inverted "V" markings.

Family: Palinuridae

Scientific name: Panulirus palyphagus

Common name: Mud spiny lobster

Local name: Shewar

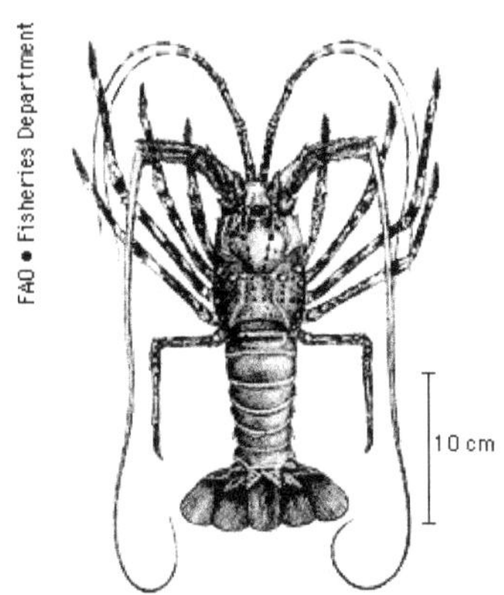

Field identification characteristics:

Broad antennular plate bearing a single pair of principal spines, antennules very long, about 1 ½ times body length, abdominal segments without transverse grooves. Legs 1 to 4 without pincers.

Colour dull greenish, abdominal segments each with a distinct transverse band of white across posterior margin. Antennules broad-banded, legs irregularly blotched creamy white.

Family: Palaemonidae

Scientific name: Nematopalemon

Common name: Spider prawn

Local name: Sungot

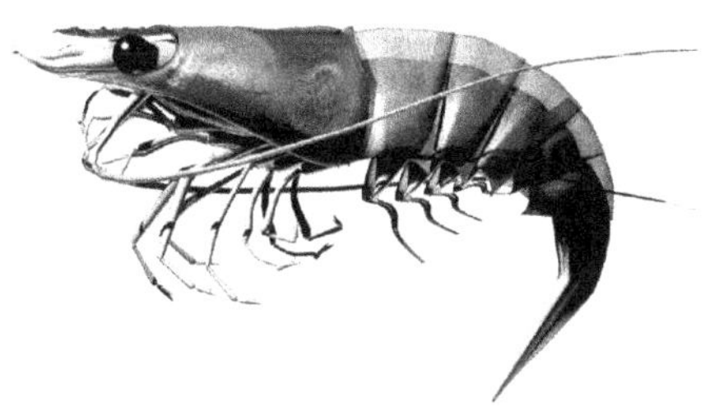

Field identification characteristics:

Rostrum long and slender, with an elevated basal crest of 4 to 7 teeth over the eye, and distal part of dorsal; margin toothless save for a small sub distal tooth, ventral margin to 6 teeth.

Colour whitish translucent with distal part of rostrum dark reddish brown, some brown Colour on antennae, anteanules and distal margins of uropods and telson, reddish brown spot on bases of uropods.

The following table shows the fish species diversity in Goa waters with common and local name.

Table-3.4 : Fish Species Diversity in Goa's Waters

Sl.No.	Common Name (English)	Local Name (Konkani)
1.	Sharks	Watu/Mori
2.	Skates	Kurscho/Mori
3.	Stingrays	Vaghole
4.	Eagle Rays	Shewny/Vaghole
5.	Catfish	Sangot
6.	Pompanos, Scads, Yellow tails	Hare
7.	Wolf-herrings	Karli
8.	Herrings, Shads, Sardines, Pel	Tarlo, Unnag, Pedwe
9.	Silver biddies	Shetki
10.	Bombay Ducks	Bombil
11.	Marlins, Sailfishes	
12.	Silver bellies	Kampi
13.	Snappers	Tamso/Tamus
14.	Goatfishes	Rane
15.	Threadfins	Dara/Rawas
16.	Croakers	Fadki
17.	Mackerels, Tunas, Albacores	Bangade, Wisvan
18.	Silver pomfret	Paplet
19.	Barracudas	Tonki
20.	Lizard fishes	Narle
21.	Engraved catfish	Sangot
22.	Thinspine sea catfish	Gaggor-Sangot
23.	Giant catfish	Sangale Sangot
24.	Soldier catfish	Sangale Sangot
25.	Indian ariomma/drift fish	Oujanje
26.	Shrimps Scads	Buranto
27.	Clef belly trevally	Kampi
28.	Yellow scad	Hare/Conkar
29.	Largenose trevally	Conkar
30.	Golden trevally	Conkar
31.	Banded scad	Conkar
32.	Indian scad	Banguli Tonki

Contd...

Table-3.4Contd..

33.	Torpedo scad	Haree/Daine
34.	Indian pompano	Conkar
35.	Barramundi	Channok/gur
36.	Thornycheck grouper	Gobro
37.	Dorab wolf-herrings	Karli
38.	Whitefin wolf-herrings	Karli
39.	Chacunda gizzard shad	Gibber
40.	Tardore	Patshali
41.	Toli shad	Pede
42.	Indian oil Sardinella	Tarlo
43.	Indian spiny turbot	Lep
44.	Malabar tonguesole	Lep
45.	Speckled tonguesole	Lep
46.	Oriental sole	Lep
47.	Golden anchovy	Kapsale
48.	Commersons anchovy	Motialli
49.	Moustached thryssa	Onang
50.	Orange mouth thryssa	Onang
51.	Saddle grunt	Korkoro
52.	False trevally	Saundale
53.	Orangefin ponyfish	Kampi
54.	Pugnose ponyfish	Kampi
55.	Whipfin silver biddy	Shatuk
56.	Spangled emperor	Tamso
57.	Mangrove red snapper	Tamso
58.	Malabar blood snapper	Palu
59.	John's snapper	Palu
60.	Flathead mullet	Shevto
61.	Indian pike conger	Toyi
62.	Goldband goatfish	Rane
63.	Striped goatfish	Rane
64.	Japanese threadfin bream	Rane
65.	Fourfinger threadfin	Rawas
66.	Sevenfinger threadfin	Rawas
67.	Indian threadfin	Dara
68.	Paradise threadfin	Runcheche
69.	Striped threadfin	Runcheche
70.	Blackspot threadfin	Rawans
71.	Moontail bullseye	Rane
72.	Goatee croaker	Dodiaro
73.	Belangers croaker	Dodiaro
74.	Blotched croaker	Dodiaro
75.	Spotted croaker	Dodiaro
76.	Bronze croaker	Dodiaro/Fadke
77.	Frigate tuna	Bugde

Contd...

Table-3.4 Contd..

78.	Kawakawa	Bugde
79.	Skipjack	Bugde
80.	Yellow tuna	Bugde
81.	Bigeye tuna	Rogtado
82.	Longtail tuna	Rogtado
83.	Indian mackerel	Bangdo
84.	Sailfish	Tarso Masso
85.	Black marlin	Tarso Masso
86.	Narrow barred Spanish mackerel	Wiswan
87.	Indo pacific king mackerel	Wiswan
88.	Pickhandle barracuda	Tonki
89.	Obtuse barracuda	Tonki/Tensori
90.	Black pomfret	Surgunteo/halva
91.	Silver pomfret	Paplet
92.	Chinese silver pomfret	Kombo
93.	Greater lizardfish	Narle
94.	Brushtooth lizardfish	Bakalau
95.	Savalai hairtail	Ballo
96.	Largehead hairtail	Ballo
97.	Jarbua terapon	Korkoro
98.	Largescaled terapon	Korkoro
99.	African flyingfish	
100.	Viviparus halfbeak	Tonki
101.	Cobia	Modso

Sl.No. SHRIMPS	Common Name (English)	Local Name (Konkani)
1.	Jinga shrimp	Sungot
2.	Speckled shrimp	Sungot
3.	Giant tiger prawn	Vagee
4.	Green tiger prawn	Sungot
5.	Hunter shrimp	Sungot
6.	Spider prawn	Sungot
7.	Mud spiny lobster	Shewar
8.	Flathead locust lobster	Sheword
9.	Christian crab	Kurli
10.	Blue swimming crab	Kurli
11.	Three spot swimming crab	Kurli
12.	Mud crab	Khadpi/kurli
13.	Indian squid	Manki
14.	Octopus	Manki
15.	Pharaoh cuttlefish	Bebbo
16.	Spineless cuttlefish	Bebbo
17.	Venus shell	Dhave Khubey
18.	Venus shell	Khubey
19.	Venus shell	Tisrio

Source : Claude Alwares (2002) Fish Curry and Rice.

From the above table attempt has been made to understand the physical attributes of some of the known species with the help of colourful pictures display. Catch (Verekar, 2000) Mackerel, locally called Bangada, is the most preferred fish of the people of Goa.

3.8 DEVELOPMENT OF INLAND FISHERY RESOURCES

Inland fishery resources are an important aspect of study of fishery resources and their marketing in the study area. Inland fishery resources have been classified into three broad categories –

1) Rivers, Estuaries, Mangrove, Brackish Water and Khazan Lands,
2) Fresh Water Lakes, Reservoirs, Canals and Other impoundments,
3) Aquaculture.

1) Rivers, Estuaries, Mangrove, Brackish Water and Khazan Lands

The study area has unique water features, that is, riverine system in the form of river branches, canals creaks and tributaries about a length of 555 kms, of which 250 kilometers of long net work of riverine water ways are navigable. The nine rivers flowing in Goa are Terekhol, Chapora, Baga, Mondovi, Zuari, Sal, Saleri, Talpona and Galgibag of which Zuari and Mandovi are the major ones (Subramanian 1994). Most of these rivers are originating from the western ghats and meeting the Arabian sea in the west.

The total estuarine area in Goa is 13,157 hectares. Mandovi and Zuari are the two major estuaries known as life lines of Goa, which are inter-connected by cumbharjua canal 14 kilometer and 11 kilometer away from the mouth of the respective estuaries. Both the estuaries open into the sea at Dona Paula beach. The salinity and tidal effects are very much pronounced, except in June-September of the year, when there is heavy inflow of fresh water into the Sea. Mondovi estuary is five kilometers wide at the mouth region, where as Zuari is about seven kilometers wide. The estuary estuarine bed of the Mondovi river is mostly sandy or muddy, where as, Zuari is mostly rocky (Subramanian, 1994).

Subramanian (1994) reported that the estuarine sector of Goa is divided into 3 zones based on the salinity gradient, substratum and wave action. These are (1) Euryheline Zone towards the river mouth with high salinity (30-35 ppt), strong wave action and rocky or sandy substratum. (2) Polyhaline Zone with moderate salinity (18-30 ppt) less wave action, sandy or clayey substratum and mangrove vegetation and (3) Mesohaline Zone in the interior region with low salinity (5 to 18 ppt) and having salty clay substratum.

Over major parts of the estuaries in the tidal Zone, traditional fishing is carried out. Such fishing is permitted under special legislation under Daman and Diu fisheries rules 1981 in 600 hectare area.

Khazan lands - There are about 18,500 hectares of Khazan lands in Goa along the banks of which around 3500 hectares. It is totally marshy and can be gainfully developed for prawn farming/pisci culture. There are around 400 tidal shell fish / prawn filtration farms along the banks/creeks and rivulets bordering on the rivers in the tidal Zone. Around 199 hectares area has been developed for prawns farming during the last five years. The yield of prawns varies from 1 to 2 tonnes per hectare per crop. Certain khazan areas are situated very close to estuary or its tributaries are demarcated for setting up a permanent fish farm. This area is flooded with saline water, and receives wild stock of all varieties of fish and shrimp during May - June period and harvesting of marketable size of fish is a continuous process (Dhandar and Subramanian, 1998). Shrimps, Multets, lady fish, pearl spot, Cat fish - estuarine crabs and Molluses form the major fisheries from the Khazan lands.

Goa is having about 2000 hectares of Mangrove area, which is associated with riverine estuaries, especially Mondovi and Zuari and Cumbharjua canal. Many mud flats are also formed on either side of the river canal.

The study area is also characterized by Brackish water bodies around 3500 hectares. About 80 hectares of brackish water area in the state has been brought under developed scientific shrimp culture with subsidy support from BFDA.

Another 19 hectare area with subsidy and support from MPEDA has also been similarly developed. About 60 percent of the marshy area, that is, 2000 hectare is estimated to be fit for aquaculture. The pilot shrimp seed hatchery project operated since 1992 under BFDA (Brackish Water Fishery Development Agency) is catering to the needs of supply to the prawn farmers of the region.

2) Fresh Water Lakes, Reservoirs, Canals & Other impoundments

The fresh water resources of Goa are comparatively very limited in the form of 3300 hectares spread in the area of Selaulim reservoir in South Goa, and 253 hectares at Anjunem dam in the North Goa. Mayem is an important water lake known for its aesthetic beauty of seven hectares located in Bichalim taluka. There are 283 fresh water badles, and the total fresh water area is about 3962 hectares. In the study area 600 hectors of area has been brought under traditional fish farming. There are 60 licenced prawns farms concentrated in various parts of Northern and Southern parts of Goa. It has been estimated that the approximate products for culture fisheries is about 3526 tons annually.

The following table shows the potentiality of Inland Fishery Resources of Goa.

Table-3.5 : Below shows the Area of Inland Fishery Resources of Goa

SI. No.	States	Rivers and Canals(kms)	Reservoirs (lakh Ha)	Tanks and ponds (lakh Ha)	Flood lakes and derelict (lakh Ha)	Blackish water (lakh Ha)	Total water (lakh Ha)
1	Goa	210	0.6	0.015	0.0016	0.0033	0.62
2	Total	210	0.6	0.015	0.0016	0.0033	0.62

Source : Directorate of Fisheries (2007-08).

Table-3.6 :Culture Fisheries

Number of prawn hatcheries	1
Number of licensed prawn farm	60
Area covered under science farm	176.31 ha
Area covered under traditional farming	600 ha
Number of fresh water fish seed hatchery	1
Number of fresh water bodies	283
Total fresh water area	3962.00
Approximate products for culture fisheries	3526 tones
Demonstration farm	1

Source : Directorate of Fisheries (2007-08).

The above tables depict the area covered by Inland fisheries in the study area. Rivers and canals constitute 210 kms, Reservoirs 0.60 lakh hectares, tanks and ponds 0.015 lakh hectares, flood plains 0.0016 lakh hectares, Brackish water 0.0033 lakh hectares and total water bodies concentrated in 0.62 lakh hectares. These areas are very useful and productive in nature, stimulating the growth of fresh water culture fisheries. The important lakes in Goa are Mayem, Curtorium and Cacoda, Benaulin, Carambolim and perennial ponds and bhandaras. There are about 400 storage tanks and 1500 minor irrigation wells in Goa. There are two major canals at Khandepar and Paroda. About 200 hectares of old and discarded nine rejects pits are available, and these are having perennial water source with a water depth of 5 to 30 meters and these can be considered for fishing operation (Subramanian, 1994).

The major inland fishes of Goa comprise mullets, pearl spot, ladyfish, betki, snappers, ambasiss species, croaker, carangids and Indian salmon, etc. The crustaceans consists of shrimps/prawns, crab and molluses consist of Green Mussels, Oyesters and Clams..

Production of Inland Fishery Resources

The production of inland fisheries has increased gradually over period of long 30 years with a small beginning of 1508 tons of fish production in the year 1972. During this time, the conditions were not much developed in the state resulting into less catch (fish catch).

The following table shows the yearwise Inland fish production during the period from 1972-2009.

Table-3.7 : Inland Fish Production During the Period from 1972-2009

Year	Production in M.Tons	(%) Growth Rate
1972	1505	–
1973	1212	-24.17
1983	1413	14.2
1985	1507	6.36
1987	1494	-0.86
1990	3046	50.95
1993	3053	0.2
1995	3562	14.2
1998	3474	-2.5
2000	3537	1.7
2001	3713	4.7
2002	3147	-17.9
2003	4284	26.5
2004	4397	2.5
2005	4184	-4.8
2006	4131	-1.5
2007	3070	-34.5
2008	3078	0.2
2009	3283	6.2

Source: Directorate of Fisheries, Govt. of Goa (2009-10).

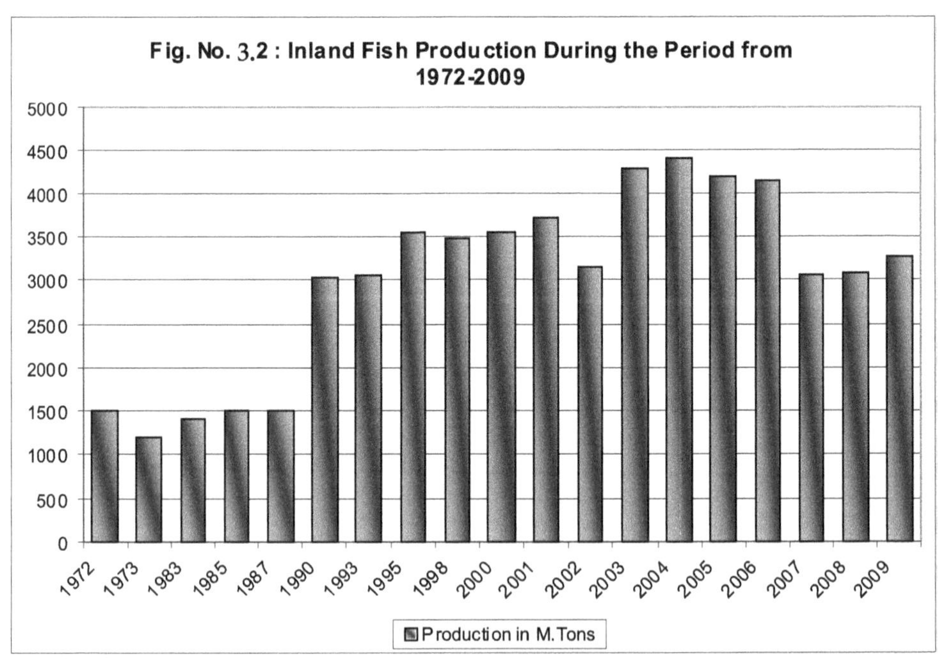

From the above table it appears that the production in Inland fish has been increasing consistently over the years, except during 1972-73 where the production was 1505 tons, 1212 tons respectively. It was a small beginning. After 1983 there is drastic change in Inland fish production, which has increased from 1413 tons to 3046 tons in 1990 and it further went up to 4397 in the year 2004. This was a remarkable growth in respect of Inland fish production, and it attracted a good value of Rs. 2211 lakhs in that year. The trend shows continuous fluctuations in the growth and decline of Inland fisheries. The best period was from 1987 to 1990 as there was 50.95% of growth, i.e., production increased from 1494 in 1987 to 3046 in 1990. It has really been a positive indicator of development of Inland fisheries in Goa.

There after conditions were highly favourable and quite encouraging, i.e., lakes, ponds, tanks, reservoirs and other fresh water bodies were brought under fisheries culture under the direction of BFDA, MPEDA and Directorate of Fisheries, Govt. of Goa. Inland fish production was somewhat stable during the period 2000-2006, which was in the range of 3500 tons to 4131 tons with minimal decline. But in the very next year 2007, the production has gone down rapidly to 34.5%, that is, to about 3070 tons from 4131 in the year 2006. This was a huge setback. The actual problem associated with Inland fisheries inspite of commendable support from the government and private agencies needs to be identified. And now the scenario has been changing quietly. There has been minimal increase in Inland fish, that is, from 3078 tons to 3283 tons in 2008-2009 respectively, which displays slight recovery in the study area @ 6.2% annually during that period.

The following table shows the trend of Inland fish production and significant values.

Table-3.8 : Trend of Inland Fish Production and Values (2003-2009)

Year	Production in M.Tons	Rate of Growth %	Values in Lakhs
2003	4284	-	-
2004	4397	2.5	2211
2005	4194	-4.8	1989
2006	4131	-1.5	2255
2007	3070	-34.5	1856
2008	3078	0.2	2047
2009	3283	6.2	2046

Source: Directorate of Fisheries (2009-10).

The above table and compound bar graph describe the trend of Inland fish production and their significant economic values.

The major Inland fishes of Goa consist of mullets, pearl spot, ladyfish, ambasis species, croakers, betki, snappers and Indian salmon. The other crustaceans are variety of prawns estuarine crab and molluses, which consist of green mussel, oyster and clams. They command good price in the local and international market. The value of Inland fish has been stagnated in the range of Rs. 2211 lakhs to roughly Rs. 2000 lakh during 2004 to 2009 period.

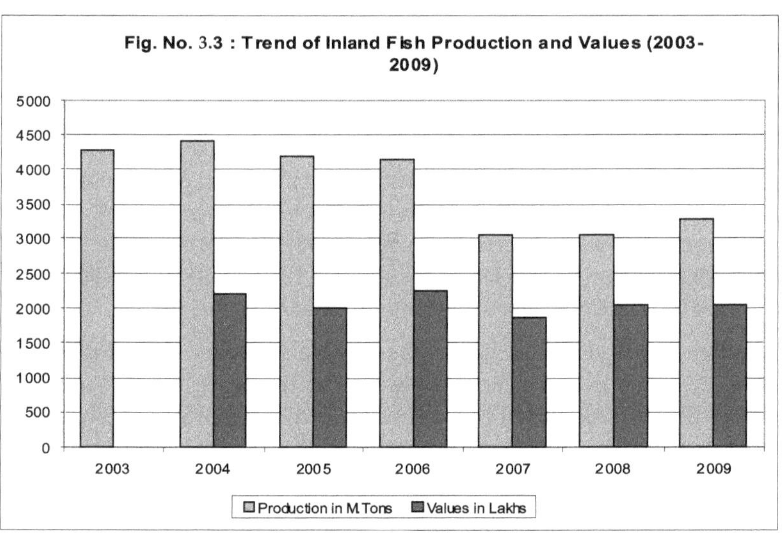

The following table shows specie wise quantity of Inland Fish catch from 2002 to 2009 in M.Tons.

The above table displays variety wise (species variety) of Inland fish catch in the study area from 2002 to 2009. It is a remarkable feature of Inland water of Goa that it produces a variety of fresh water species, brackish water fish and skimps/prawns over the years. Important species are prawns, big, small and medium, lady fish, mallets, gerres, dutianus, catfish, ambasis, crabs, black water clamps, false clamps oysters, balle, green clamps and others.

The total production of Inland fish has been 3078 tons as on march 2009, compared to 4397 tons in the year 2004 valued at Rs. 2211 lakhs. This was the highest in respect of Inland fish production in Goa, followed by 4194, 4131 in the year 2005, 2006 and they were valued at Rs. 1989, Rs. 2255 lakhs respectively. The major species production is confined to small prawns mallets, black water clamps, catfish, pearl spot, crabs and false clamps as compared to other species. Ladyfish, green clamps, ambasis, betki, lutianus, etc. are caught in small quantities within the range of 100 metric tons annually. Small size prawns are the outstanding component in Inland fish production, and their production is in the range of 1000 M.tons to 600 M.tons annually during the period from 2002 to 2009, followed by mullets and black water clamps. All these are highly economically significant, fetching good income prawns of big size and medium size which attract very good high price from the point of caring higher income, but unfortunately there are less number in years because these species are out of reach of common man as he cannot afford to buy such costly prawns because of his low purchasing power. Thus resulting into less production, that is, in the range of 20-100 tons (big size) prawns.

Lady fish has been caught in small quantities in the range of 20 to 80 metric tons annually and it is known for very good taste. The other important species are pearl spot, oysters, green clams, crabs and ambasis extracted in different small quantities over the years.

Table-3.9 :Specie wise Quantity of Inland Fish Catch from 2002 to 2009

Sl.No.	Name of Fish	2002	2003	2004	2005	2006	2007	2008	2009
1	Prawn: big medium small	20	62	64	46	48	15	11	99
		262	137	150	111	158	77	41	460
		958	977	987	716	979	811	571	618
2	Lady fish	77	70	74	80	76	38	26	24
3	Mullets	144	187	186	207	196	233	304	267
4	Genres		59	67	65	80	70	81	69
5	Lutranus	28	23	35	45	37	5	7	4
6	Cat fish	254	219	205	210	196	111	115	103
7	Anchovy	31	22	31	34	35	3	1	4
8	Pearl spot	59	62	59	71	64	116	140	122
9	Betki	6	4	5	7	4	1	1	0.3
10	Milk fish	-	-	-	1	-	30	26	23
11	Megalops	1	1	1	1	-	4	1	14
12	Scatophagus	31	26	31	31	33	36	37	32
13	Ambasis	111	94	83	92	88	9	27	34
14	Crabs	157	133	143	161	138	116	116	108
15	Black water clamps	60	723	756	716	718	604	307	237
16	False clamps	293	194	179	163	155	70	360	113
17	Oysters	19	81	380	184	167	1	1	1
18	Balle	5	-	-	-	4	-	1	0
19	Green clamps	46	613	364	464	358	88	113	0
20	Miscellaneous	524	583	594	765	607	625	791	703
Q	Total	3147	4284	4397	4194	4131	3070	3078	3283
V	In Lakh Rs.	-	-	2211	1989	2255	1856	2047	2083

Source: Direction of Fisheries, Govt. of Goa (2009-10).

The following table shows the Demographic Profile of Inland Fishery Resource Development.

Table-3.10 : Goa – Inland Fishing – Active Fishermen Population

Inland Fish Landing Centres	Fishermen Population
Siolim	100
Badem	136
Colrale	188
Camerlium	380
Aldona	112
Pomburpa	350
Vritona	3995
St. Estevam	415
Cumbharjua	525
Diwar	225
Chorao	130
Mandur	150
Goa Velha	230
Neura	78
Agaciam	210
Curca	23
Bandora	80
Shiroda	120
Borim	85
Durbhat	350
Madkai	370
Colomb	110
Patne	140
Kiniebag	245
Chiclaim	80
Ambelim	71
Assolua	42
Bambolim	160
Salvor demunds	46
Chodan	113
Total	9159

Source : Directorate of Fisheries (2009-10).

GOA – POPULATION OF INLAND FISHING CENTERS

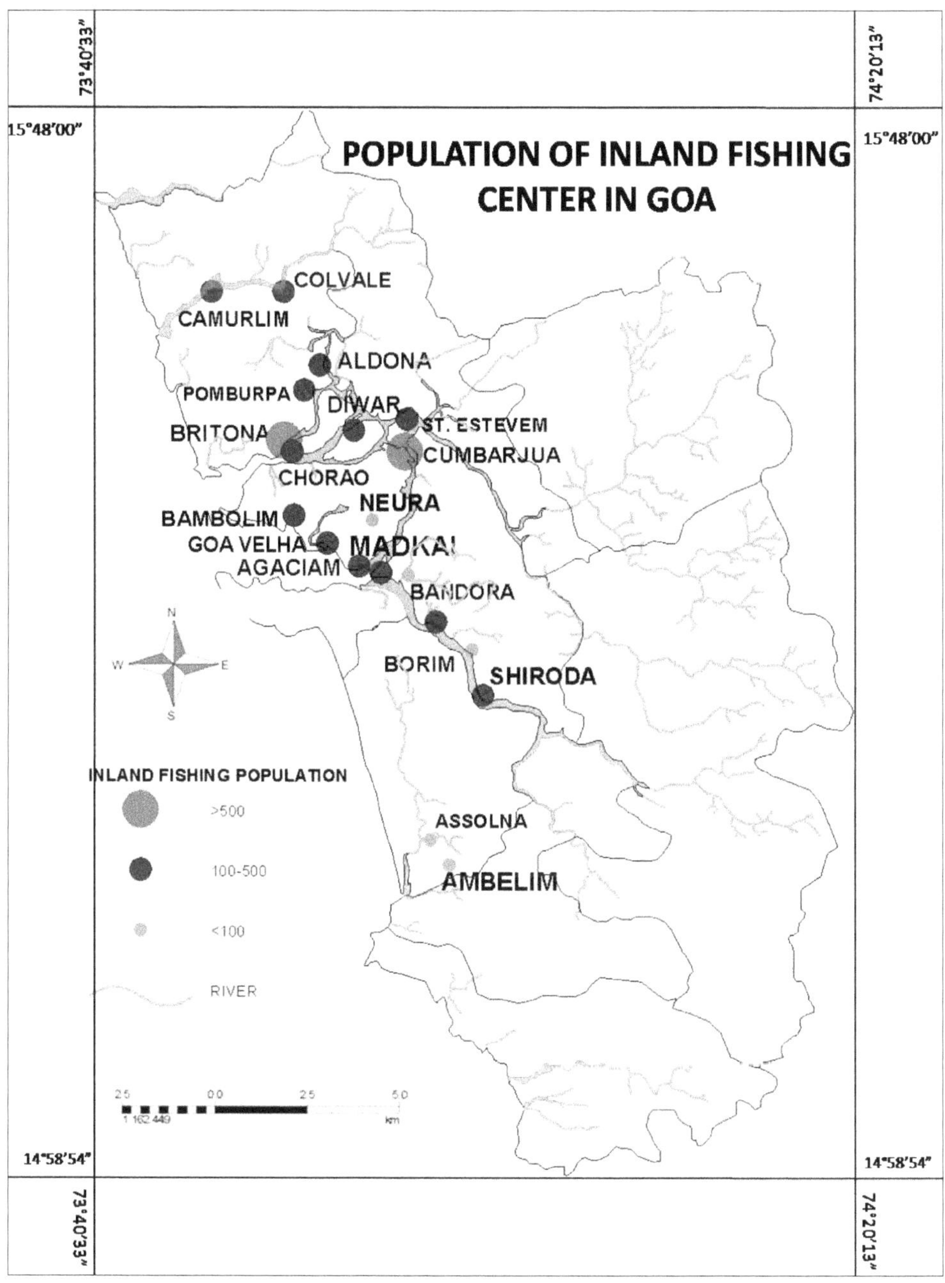

Map-3.1

Inland fishery sector has 9159 active fishermen population in the study area. From the table given it has been observed that various active fishermen in varying proportion are engaged in inland fishing activities in different parts of Goa. In all, there are 9159 active full time fishermen. Britonna from Bardez taluka records the highest number of fishermen population, i.e., 3995 in the study area, whereas St. Estevam and Cumbharjua from Tiswadi taluka have moderate size, i.e., 415, 525 fishermen population engaged in fishing activities, while the lowest is recorded by Curca from Tiswadi taluk with just 23 fishermen as its smaller fishing area. Hence the scope is limited.

3.9 AQUACULTURE

Introduction

India is endowed with a long coastline, and hence it offers ample scope for large exploitation of marine resources. Till a few years back, fishermen in India were only involved in traditional marine fishing. In the 70's fishermen started concentrating on catching prawns, more commonly known as "prawns". It is basically due to high profitable return on account of their high export value. Brackish water prawns faming started in a big way during 91-94, especially in the coastal districts of Andhra Pradesh and Tamil Nadu. It is considered to be the most promising part of aquaculture in India in recent times.

The prawn farming has now been regulated with the establishment of Aquaculture Authority of India. It is as per the direction of the Supreme Court of India for issuing licenses and overall supervision. It is commonly said that after green and white revolution in India, it is the time for blue revolution to exploit the huge potential in the fisheries sector. Prawns are called the "pinkish gold" of the sea because of its universal appeal, unique taste, high unit value and increasing demand in the market.

Prawn farming, in Goa, is considered to be the most lucrative enterprise due to high market demand price and persistent global demand. Prawn farming development also holds immense employment potential to the local people. It provides direct employment to at lest 2 persons per one hectare cultured area. It also provides indirect employment to the extent of 1100 men per day for one hectare during construction, input supplies, handling of material, harvesting, etc.

Although the cultured prawn is a nutritious food item, the domestic market for shrimp is quite negligible. Only peeled and second rate prawn is available in the domestic market. It is due to the fact that prawns are exported to the international market where it fetches high price. Culture of prawn, thus, helps the state or country to earn foreign exchange, which is very precious to a developing county like India.

Culture of prawn, under brackish water environment, apart from providing subsidiary income and employment to the farmers, also contributes to augment foreign exchange earning and to accumulate foreign exchange reserves. These advantages, therefore, made prawn farming an export oriented, priority allied activity in the process of India's developmental planning. Brackish water shrimp culture is considered a promising activity to provide employment and subsidiary income to the rural people.

Relative Advantages of Cultured Prawns over Captured Fisheries

Scientific aquaculture can ensure steady and regular production of fish, which may be increased considerably with planned input of modern technology, just like scientific agriculture. Of all thee aquaculture products, prawn is considered to be the most important by virtue of its taste, high unit valve, persistent demand and universal appeal among the wealthier nations of the world.

Cultured prawn production has some advantages over capture prawn fishing. First of all, shrimp catch is seasonal and unreliable even in tropical waters and may fluctuate from year to year. Cultured prawn production, on the other hand, can be planned in advance at least in the tropics and can be run continuously. This improves the efficiency of processors, and may reduce the cost of exporting and importing.

Prawn catch from wild source is determined by the geographical and seasonal conditions. But shrimp farmers can select more flexible ones from wide range of cultural species and can reduce the size required according to the consumer demand.

Cultured prawn is capable to withstand transport and can be kept alive for longer than time than the captured prawns.

Factors influencing Prawn Farming

1. Availability of suitable land environment

One of the major factors influencing prawn farming activities is the availability of suitable soil and water. The site should contain soft bottom soil, mixed soil comprising of clay, sand and silt to ensure good water bearing capacity as well as production of natural food organism on which prawns could feed and grow. The tidal water, that is the basic water source for better growth of prawns, is 25-35 percent, the PH should be 5-7. The meteorological parameters such as rainfall and its distribution, wind speed and direction, humidity, sunshine and evaporation rate also affect the production of prawns. In Goa, around 600 hectares of land is scientifically developed for prawn farming.

2. Profitability

Any enterprise depends upon profitability, and prawn farming is no exception. Short direction of crop, on investment and fast expanding attractive world and local market make prawns farming lucrative and profitable activity. People go in for prawn farming because they get a lot of profit depending on the crop. They put 10000 seeds and if the crop is good they get a profit of Rs. 100000. But if the crop fails the investors lose everything. Big money depends upon the crop.

3. Suitable Climate

Moderate temperature is required for the prawn aquaculture. The optimum range temperature for the Black tiger prawns is between 28′c–30′c, the metabolism reduces and so does the active behavior and growth rate.

4. Brackish Waters

Goa being a coastal belt, 104 kms we have a lot of water estuaries; this makes it suitable for conducting prawns farming activity. There are about 171.155 hectares of total area of prawns farming fields in various part of north and south Goa.

5. Availability of Labour

Labour for prawns farming is easily and cheaply available in Goa. Both skilled and unskilled labour is easily found in Goa. Labour has been coming from neighboring states to carry out the necessary operations.

6. Government Support

The Government of Goa conducts courses in Goa on aquaculture. They have built a demonstration farm in Ela Dauji, old Goa, especially to conduct courses in prawns farming. Both Government of Goa and MPEDA offer subsidies to licensed farmers. BFDA is also playing a vital role in development aquaculture.

7. Market

Since the scarcity of sea catch is increasing year by year, brackish water prawns culture is increasing. There is a huge demand for prawns in local market and world market. Prawns culture contributes a huge amount of foreign exchange to the Goa's economy.

Varieties of Prawns

The important varieties of prawns are:

1. Black Tiger Prawns

The penaeid shrimps, the black tiger prawns, which are popularly known as "jumbo tiger prawns" (P. Monodon) are the most popular species for the culture because of its high growth potential and price. It attains maximum size of 330mm in natural habitat, 240 mm in pond growing conditions. Tiger prawn is highly relished by the gourmets all over the world in general and Goa in particular.

2. Indian White Shrimps

The second important species is the "Indian white shrimp" (P. Indicus), which is versatile, but has less growth potential than P.Monodon. Its size ranges from 140-170 mm. Shrimps of this type are cultured extensively and almost exclusively under the brackish water aquaculture in India.

Techniques of Prawn Farming

1. Traditional

In the traditional system of prawn farming, prawn seeds along with the seeds are trapped during the high tides when the ponds get inundated. In this process many predators also get entry, and they all co-exist in the same pond. Prawns are often harvested before they are grown to optimum size. Owing to indiscriminate stocking of both desirable and undesirable varieties of prawns, a fish etc., the production of prawn under this system is normally unpredictable and often low in quality because of predation by fishes, etc. The average production from traditional system is below 0.5 tons/ha/annum.

2. Extensive

The extensive system of prawn farming is an improved method of traditional farming involving construction of new ponds, ranging from 1-5 ha in suitable selected area. Selective stocking with fast growing shrimps seeds at a comparatively lower density ranging from a few thousands to 1,00,000 seeds per ha with supplementary feeding. The average production under this system normally ranges from 1-1.5 tons/ ha/crop.

3. Semi Intensive

The semi intensive system of prawns farming involves construction of ponds, ranging from 0.2-0.5 ha in size, selective stocking with fast growing hatchery seeds at a comparatively high density ranging from 1.3 lakh/ha maintenance of water quality by exchanging 10-20% daily aerating of the pond with air blowers/puddle wheels and feeding the prawns with nutritious feed. The average production in this system ranges from 4 to 5 tons/ha/crop in 4-5 months.

4. Intensive

The intensive system of prawns farming involves construction of concrete ponds of 0.3-0.1 ha in size, selective stocking quality prawn seeds procured from hatcheries at a density ranging from 5-10 lakh ha, maintaining water quality by exchanging over 30 percent a day, aeration of the pond. With mechanized aerators and feeding the prawns with nutritionally well balanced high energy feed, the production from the intensive system ranges from 10-20 tons/ha/crop.

Distribution and Production of Prawns Farms/Shrimps in Goa

North Goa

The prawn/shrimps farming activities are widely spread in different proportions in south Goa. They are largely concentrated in south Goa as compared to North Goa.

The distribution of prawns farms in north Goa is mainly confined to Chorao, Betim, Durbhat, Tuem, Paliem, Colvale, Mavcel, comulim and Divar. Most of the fields are situated along the banks of the river, Mondow Chapora and their tributaries. The total area covered by prawn farms in North Goa is about 81.72 hectare as on 2008-09 records, out of which Chorao ranks first in having maximum area, i.e., 14.91 hectares, production and output values are 41074 M.tons and Rs. 68.8 lakes, followed by Diwar 16.00 hectares of area producing about 44.8 M.tons of prawns valued at Rs. 73.8 lakh in 2008.

Paliem ranks third in respect of area with 14.00 hectares, production and out values, as they are experiencing highly favourable conditions for development of prawn farming activities. Each hectare produces approximately 2.81 metric tons of shrimps, and they are valued at Rs. 164847.51 per ton as per 2008-09 prices. Major part of the produce has been exported. The total production of prawn in north Goa is about 231.09 M.tons valued at Rs. 3.80 crores. There are, in all, 35 prawn farms located in 16 villages of north Goa, where Bandora, Tiviul, Madaki, Mavcel are noted for less water spread area for the development of prawns farming activities. Hence, they have less productivity, i.e., Tiuim 2.0 M.tons and Marcel 1.59 M.tons, Madkai 5.46 M.tons and Marcel 0.89 M.tons respectively. Thus these areas earn less amount of income compared to other areas.

The following table shows the Distribution and Production of prawn farms in North Goa, 2008-09 outside and within CRZ

Table-3.11 : Distribution and Production of Prawn Farms in North Goa, 2008-09

Sl.No.	Location	Total area in hectare	Production per hectors 2.81 in M.tons	Value in crores
1	Chorao	14.91	41.74	0.68
2	Betim	2.8	7.84	0.12
3	Durbhat	2.5	7	0.11
4	Tuem	4.72	13.21	0.21
5	Palieur	14	39.2	0.64
6	Chopdem	1.69	4.7	0.07
7	Siolim	1	2.8	0.04
8	Pilerue	10.2	28.56	0.47
9	Coluale	4.11	11.5	0.18
10	Talanium	2.17	6.07	0.1
11	Marcel	0.32	0.89	0.01
12	Diwar	16	44.8	0.73
13	Camurlim	2.68	6.66	0.1
14	Madkai	1.95	5.46	0.09
15	Tivim	0.715	2	0.03
16	Bandora	0.57	1.59	0.02
	Total	**81.72**	**231.09**	**380**

Source: Directorate of Fisheries, Government of Goa (2008-09).

Note: 2008 - 623 tons, value Rs 10,27,00,000. Per ton - 164847.571

Distribution and Production of Prawn Farms in South Goa 2008-09 (Outside and Within CRZ)

The total area under prawn farming is comparatively higher in south Goa than north Goa. There are 50 prawn/shrimp farms concentrated in 19 village of south Goa, having 139.66 hectors water spread area for prawns culture, such as Cartorim, Chinehinim, Quolosim, Sanelie, Panchawadi, Benaulia, Deussua, Carmona, Loliem, Cauelasim, Ambelim, Poinguinim, Chiclaim, Cortalim, Navelim, Silim velim and Telauliar.

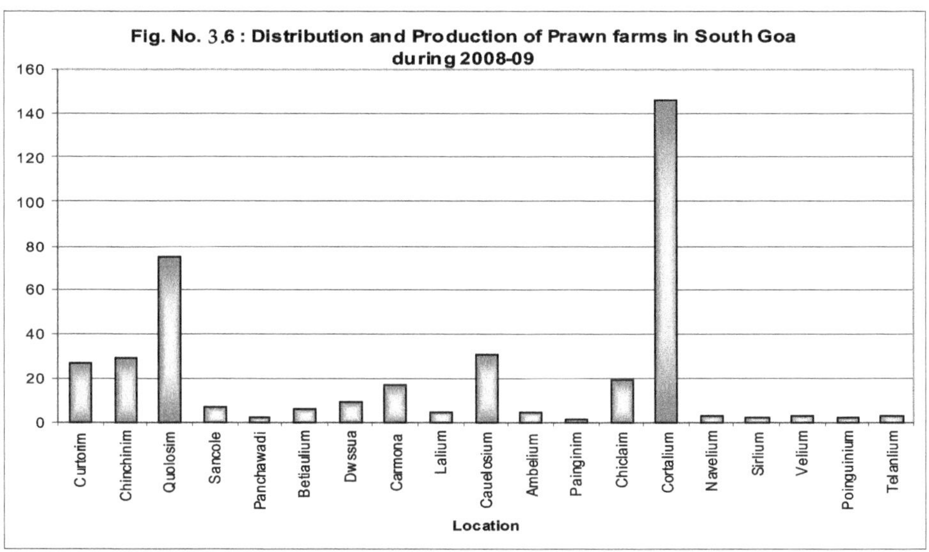

Cortalim ranks first in both total area as well as total production. It has 52.29 hectares water spread Area and production about 146.4 M.tons valued at Rs. 2.41 crores in 2008-09, followed by Quolosim 29.71 hectors, 74.78 M.tons of prawns production and valued at Rs. 1.23 crores.

Chinchinium ranks fourth in area, production and output values. It has 10.42 hectare area and produced about 29.17 M.tons prawns, which are valued at Rs. 0.48 crores.

The following table shows the Distribution and Production of Prawn farms in south Goa during 2008-09. (Outside and within CRZ)

Table-3.12 : Distribution and Production of Prawn Farms in South Goa during 2008-09

Sl.No.	Location	Total area in hectare	Estimated production	Value in Rs. (crores)
1	Curtorim	9.59	26.85	0.44
2	Chinchinim	10.42	29.17	0.48
3	Quolosim	26.71	74.78	1.23
4	Sancole	2.49	6.97	0.11
5	Panchawadi	0.72	1.44	0.02
6	Benaulium	2.87	5.74	0.09
7	Dwssua	3.35	9.38	0.15
8	Carmona	5.9	16.52	0.27
9	Lalium	1.44	4	0.06
10	Cauelosium	10.9	30.52	0.5
11	Ambelium	1.65	4.62	0.07

Contd...

Table-3.12 Contd..

12	Painginim	0.4	1.12	0.01
13	Chiclaim	6.9	19.32	0.31
14	Cortalium	52.29	146.41	2.41
15	Navelium	0.91	2.54	0.04
16	Sirlium	0.7	1.96	0.03
17	Velium	1	2.8	0.04
18	Poinguinium	0.6117	1.71	0.02
19	Telanlium	0.81	2.26	0.03
	TOTAL	**139.66**	**391.91**	**6.46**

Source: Directorate of Fisheries, Government of Goa (2008-09)

From the above table it appears that the south Goa accounts for 139.66 hectors of water spread area for prawns/shrimp farming and produced about 391.91 M.tons, and its output value is estimated at 6.46 crores, considerably a good income towards the state economy of Goa. Thus in all North Goa and south Goa together about 623 M.tons of prawns is produced in the year 2008-09, and they are valued at Rs. 10.39 crores.

Benauliam, Carmona, Cavlosium, Denssus Curtoriam are also important prawn producers in south Goa. Especially Caulosim ranks third in respect of area as well as production and output values. Chiclaium is also doing well in prawn farming activities, having an area of 6.90 hectares and produced about 19.32 M.tons. The least produce is in Navelium, Sirlium, Velium, Pongainium and Telaulium on account of less area, resulting in less production and less output values in the state/south Goa.

The Distribution of prawn/shrimp farms in both north Goa and south Goa can be displayed with the help of pictorial map of Goa.

The following table shows the trend of Prawn/Shrimp farming i.e. production and output value in the study area during 2004-05 to 2008-09.

Table-3.13 : Inland Prawns Production in Goa

Sl.No.	Year	Production M.tons	% Rate of Growth	Value Output (in crores)
1	2004 - 05	847	-	10.2
2	2005 - 06	869	2.53	11.68
3	2006 - 07	1184	26.6	14.54
4	2007 - 08	903	-31.1	9.33
5	2008 - 09	623	-44.94	10.27

Source: Directorate of Fisheries, Government of Goa (2008-09).

From the about table it appears that, irrespective of increase or decline of production, Prawns invariably command very good price in the local and international market. During 2006-07, the state has earned Rs. 14.54 crores by producing 1184 M.tons of Prawns of varying size, i.e., big, medium and small size. The important

Fig. No. 3.7 : Inland Prawns Production in Goa

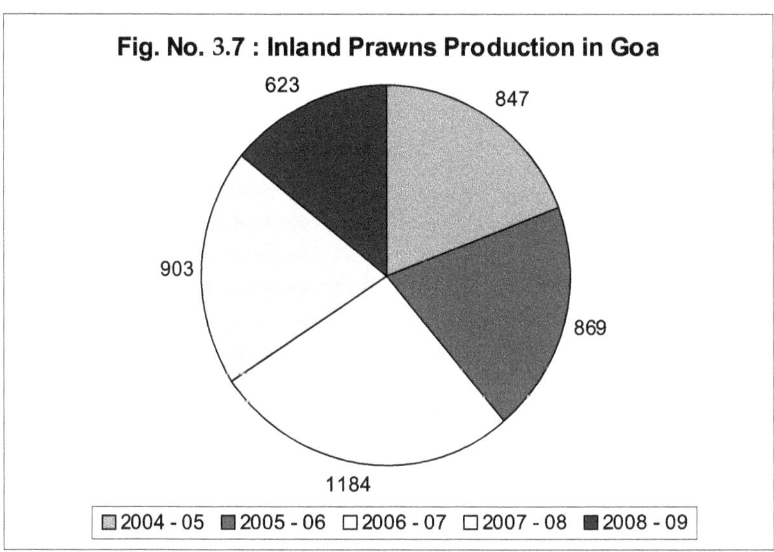

ones are Black Tiger Prawns and Indian White Shrimps, which attract good price in the market. The trend shows decline in the productivity from 2007 onwards reaching low figures, i.e., 623 M.tons almost 40 percent decline. It is a matter of serious concern. Appropriate measures need to be taken to prevent the decline in the productivity of shrimps in different parts of the study area.

3.10 DEVELOPMENT OF MARINE FISHERIES

Goa has a large scope for fisheries production, particularly, through brackish water and marine production. Marine fishing is a seasonal industry for a period of nine months. The fishing season commences from the mid of August, when the fishing ban is lifted in the study area and lasts till the end of May. Marine fisheries are closed (practically closed) during monsoon, and fisherman utilize this period for mending their nets, maintenance of the fishing vessels and boats and also preparing new nets, etc. (Directorate of Fisheries : 2007, Govt. of Goa).

Marine fishery resource of Goa comprising 104 kms of coastline (Directorate of Fisheries : 2007) characterized by innumerable creeks, bays, mangroves, swamps and coral reef, etc. It's a broken coastline, ideal for coastal navigation and development of fish landing centres (centers), which is 1.25 percent of the country's total of 8192 kms. The continental shelf area extends upto 10,000 sq.kms 100 fathoms depth. Marine fishery resources (fisheries) have developed extensively over the years due to favorable conditions. According to Parulekar (1989), the potential pelagic fishery resources for the EEZ (Exclusive Economic Zone) are 69000 tons for the shelf, and 8000 tons for the oceanic area. The sustainable pelagic yield is projected as 46,560 tons per annum. Similarly, the potential demersal resources of EEZ are estimated to be 1,12,600 tones with a sustainable yield of 67,500 tonnes per year [Dr. Subramanian (2002)]. Therefore, the total sustainable yield for both pelagic and demersal fisheries of Goa is projected to be 1,14,060 tones annually.

There are seven taluks located on the coast with 42 marine fishing villages engaged in extraction of marine fisheries. The highly productive fishable area in the sea is extended upto 20-40 fathoms and covers approximately a total area of the 2000 sq.miles. Major part of the fisherman population of 36894 is engaged in marine fishing activities to earn their livelihood (Directorate of Fisheries : 2008).

The following table displays the trends of marine fishery resource production and out put values over the years from 1960-61 to 2009-10 in the study area.

Table-3.14 : Production and Output Values – The Trend of Marine Fisheries Resources 1960-61 to 2009-10

Year	Fish Production	% of Growth	Estimate Value
in M Tones			in (Rs.) crores
1960-61	20000	-	-
1961-62	17000	-17.6	78
1970-71	36616	53.5	9
1980-81	25715	-42.3	9.85
1990-91	53179	51.6	28.5
1992-93	97014	45.1	47.5
1995-96	85418	-13.5	54
1997-98	94547	9.6	66.08
1998-99	67236	-40.6	92.21
1999-00	60075	-12.4	78.6
2000-01	64563	6.95	144.39
2001-02	69386	6.95	152.77
2002-03	67563	-2.6	133.18
2003-04	83756	19.33	155.30
2004-05	84394	0.75	171.96
2005-06	103091	1.81	271.10
2006-07	96326	-7.0	263.82
2007-08	91185	-5.6	344.46
2008-09	88771	-2.7	332.69
2009-10	80687	-10.01	346.35

Source: Directorate of Fisheries, Govt. of Goa (2009-10).

From the above table and graph, it appears that the state has exploited 20000 tons of fish in the year 1960-61. There were hardly any mechanized trawlers used. Most of the fishing activities have been carried out with the help of non-mechanized boats or traditional country crafts (wooden). In the very next year, the production of marine fisheries went down by -17.6 percent at 17000 tons annually due to decline in the catch.

After a gap of 10 years, the production of marine fisheries has increased from 17000 tons in 1961-62 to 36616 tons in 1970-71. A net 53.5 percent growth was experienced, followed by a drastic decline of -42.3 percent to 25715 tons valued at Rs. 9.85 crores during 1980-81. Production was fluctuated in the range of 20000 tons to 40000 tons during that period, wherein 282 trawlers, 36 liners and around 2205 traditional boats were used for extraction of fish from Goa's waters.

The production of fisheries saw a net growth of +51.6 percent by 1990-91 and it has gradually increased to 53179 tons – and its commercial value is around 28.50 crores. A significant growth has taken place in the study area. Infact that was real beginning of commercial exploitation of fishery resources of Goa, which enabled further large scale output of fisheries at 97014 metric tons and the value of fishery output had increased to 47.50 crores in 1992-93. A remarkable 45.1 percent growth in the fishery sector was witnessed, which could be a real boost to Goa's economy. The fishery output was greatly influenced by good catch by fishermen, and the use of 1757 mechanized boats. There was good balance between demand and supply of fish in the market centers of Goa to meet the need of local consumers, domestic tourists, internal tourists and for export trade.

Thus the production of fisheries was in the range of 80000 tons to 97000 tons till 1997-98. After this period, fishery output in Goa suddenly came down, dropped adversely to -40.6 percent from 94547 tons in 1997-98 to a meager 67236 tons in 1998-99, followed by further -12.4 percent decline in 1999-2000. Thus the fishery output in Goa has touched a low level of 60075 metric tons after a long gap of 10 years. This was greatly influenced by unfavourable weather conditions. Long dry spells, pollution, etc., led to insufficient catch resulting in to low fishery output. But there was good command over price of the fish commodity and hence value of the fish did not come down. In spite of less catch, the value of fish stood at 92.21 crores in 1998-99. It shows the significance of fish commodity commercially.

By 2001-02 onwards there was substantial increase in the fishing output and value of the fish, i.e., the state had exploited 69386 metric tons of fish valued at Rs. 152.77 crores. Subsequently, the fishery output has touched its highest 103091 metric tons, valued at Rs. 271.10 crores in the study area. This was the net outcome of appropriate measures, which were taken by Directorate of Fisheries, Govt. of Goa to increase the output; and favourable climate favoured the tremendous growth of fishery sector of Goa.

The growth is also influenced by high degree of mechanization i.e. there are 1134 trawlers and 968 mechanized canoes used for exploitation of fish. Thus, the fishery output has been increased with minor fluctuations and the output has been on an average of 90000 metric tons in the last five years. The state has exploited 80687 metric tons of marine fish in 2009-10 and the fishery output value has been around 346.35 crores, a good rise in the value of fish despite its marginal decline in the catch at present in the study area.

Specie-wise Marine Fisheries Production

The following table displays the trend of specie-wise production of marine fisheries during the period 2000-01 to 2009-10. It is quiet clear that among various

Table-3.15 : Important Species-wise Production of Marine Fish 2000-01 to 2009-10

Sl.No.	Species	2000-01	2001-02	2002-03	2003-04	2004-05	2005-06	2006-07	2007-08	2008-09	2009-10
1	Mackerels (Bangdo)	16589	14204	8103	5779	6303	12006	12244	19980	16597	15169
2	Sardines (Tarlo)	19836	21470	30951	30874	34203	28246	30558	28574	32389	23496
3	Cat Fish (Sangot)	676	2436	1585	1007	1043	1303	2586	1821	1480	2279
4	Shark Fish (Mori)	981	1211	1355	1571	1305	1716	988	1007	1019	792
5	Seer Fish (Wiswan)	1398	2746	1230	2274	3478	9556	3522	4407	3777	2295
6	Prawns (Sungtam)	2284	874	2299	6656	5586	10599	9065	8642	7458	9795
7	Pomprets (Paplet)	681	859	500	825	568	720	446	559	534	284
8	Cuttle Fish (Balle)	829	2027	832	2029	1737	73	1919	330	2201	1595
9	Tuna (Bokdo)	911	198	669	1241	609	2459	589	1916	818	1044
10	Ribbon Fish (Balle)	1430	291	2645	3178	3647	5791	5354	1368	2551	558
11	Reef Cod (Gobro)	83	699	99	769	2371	-	-	-	-	266
12	Kowalkawal (Velli)	990	553	315	268	421	854	724	226	455	395
13	Golden Anchovy(Kapsale)	82	68	129	63	9	17	30	3	57	-
14	Silver Belly (Kampi)	1533	3818	3331	2740	2489	2046	3187	1664	1688	1976
15	Soles (Lepo)	2037	1747	1682	2935	2016	2054	2339	1950	1795	1532
16	Silver Bar (Karli)	433	388	312	370	336	849	435	251	336	637
17	Crabs (Kurlio)	805	893	654	866	972	994	1092	819	1021	1349
18	Sciaenoids (Dodiaro)	2758	2451	1629	2806	2634	3211	2404	1938	2780	2207
19	Butter Fish (Soundale)	915	790	287	439	340	598	892	615	655	1353
20	Others	9312	11663	8956	11066	14327	19995	17952	15115	11160	13941
	Total	**64563**	**69386**	**67563**	**83756**	**84394**	**103087**	**96326**	**91185**	**88771**	**80687**

species sardines output is quite large, followed by mackerels (Bangdo) which is most preferred fish by the Goan consumers.

Director, Verlekar (2001), Department of fisheries, Government of Goa, shark, seer fish are produced in the range of 1000 metric tons to 5000 metric tons, followed by prawns in the range of 2000 M.tons to 8000 M.tons in the last ten years. Pomprets fish represent less productivity of 500 to 1000 tons as compared to many other species. Thus study area has greater variety of fish for consumption purpose at local as well as international level.

Table-3.16 : Share of Fishing Industry of Goa in India's Fishery Sector (1950-51 to 2009-10)

(Quantity in lakh tons)

	INDIA			GOA			
Year	**Marine Fish**	**Inland Fish**	**Total**	**Marine Fish**	**Inland Fish**	**Total**	**%**
1950-51	5.34	2.18	7.52	-	-	-	-
1960-61	8.8	2.8	11.6	0.2	-	0.2	1.72
1970-71	10.86	6.7	17.56	0.36	0.015	0.37	2.1
1980-81	15.55	6.87	24.42	0.25	0.014	0.26	1.06
1990-91	23	15.36	38.36	0.53	0.03	0.56	1.45
1995-96	27.07	22.42	49.49	0.85	0.035	0.88	1.77
2000-01	28.11	28.45	56.56	0.64	0.035	0.67	1.18
2005-06	28	38	66	1.03	0.041	1.07	1.62
2006-07	31	38	69	0.96	0.041	1	1.44
2009-10	-	-	-	0.8	0.032	0.83	-

Source: Directorate of Fisheries, Government of Goa (2009-10).

The above table displays the comparative trend of marine and inland fishery resource in India as well as in Goa during 1950-51 and 2009-10. It has been observed that the production of inland and marine fisheries in India together was about 7.52 lakh tons in 1950-51, out of which marine fish was around 5.34 lakh tons, inland fish 2.18 lakh tons. But during that time, Goa did not have substantial output of fisheries. Hence, the data was not available. Since it was under Portuguese colonial rule, the records were not maintained properly.

During 1960-61 Goa has produced about 0.25 lakh tons of marine fish compared to India's total output of 11.60 lakh tons. It has been examined that the highest output of fisheries in India is about 69.00 lakh tons as on 2006-07, while Goa has produced about 1.03 lakh tons, which accounts for 1.64 percent of total output of India in 2005-06. These are the peak stages of fishing industry of India Goa as well. High degree of mechanization and other appropriate measures undertaken by fishery authorities, i.e., Govt. of India, Goa helped in promoting sustainable growth of fishery resources, i.e., marine as well as inland.

Moderate fishery output was achieved by Goa in the year 1995-96 with 0.88 lakh tons, which accounts for 1.77 percent of India's total fishery output, i.e., 49.49 lakh tons. Further, the lowest fishery output of India was recorded in 1950-51 with 7.52 lakh tons of fish, out of which 5.34 lakh tons of marine fish and 2.18 lakh tons of inland fish was produced. On an average, the state of Goa has been constricting at 1.6 percent of fishery resource to India's total every year, which is quite low compared to other states in the country, because the state is having a small size of coastal belt with limited number of fish landing centres, etc.

The following table shows the contribution of fishing industry to Goa's economy and the share of fishery sector (industry) in the primary sector of the study area during the period 2001-02 to 2007-08.

Table-3.17 : Contribution of Fishing Industry to Goa's Economy and Its Share in the Primary Sector

Year	NSDP of Fishing Industry at Current Prices Rs. (Lakh)	Total NSDP of Goa	Share of Fishing Industry in NSDP (%)	Primary Sector Rs. (Lakh)	Share of Fishing Industry (%)
2001-02	159.7	6157.71	2.59	834.72	19.1
2002-03	169.02	7008.35	2.41	913.22	18.5
2003-04	211.82	8050.16	2.63	1102.52	19.2
2004-05	206	10039.33	2	1300.62	15.8
2005-06	330.35	11587.35	2.85	1709.14	19.3
2006-07	370.41	13273.93	2.33	2167.27	14.3
2007-08	352.22	14984.28	2.35	2566.96	13.7

Source: Economic Survey, Government of Goa (2008-09)

From the above table it appears that the fishing industry of Goa has contributed 159.70 crores to Goa's economy in 2001-02, which accounts for 2.59 percent of total NSDP of Goa and its share was attributed to 19.1 percent in the primary sector which further rose to 352.22 crores during 2007-08. A significant growth in the fishery sector, which also influenced primary sector, having 13.7 percent share of total produce in the study area. Thus fishing industry of Goa plays a vital role in socio-economic development of Goa by contributing substantially towards Net State Domestic Product through Export and Domestic trade annually, and subsidiary economic activities have grown around fish harvesting in the study area.

The following table shows the quantity and value of marine fish and marine prawns over the years from 1998-2009. The trend shows a remarkable growth in fishery output, prawns and their respective values in the study area.

Table-3.18 : Quantity and Value of Marine Fish and Marine Prawns

Qty. in M.Tons Value in Lakh (Rs.)

Year	Marine Fish		Marine Prawns		% of Value of Prawn to Total Value (Percentage)
	Qty	Value (Rs.) in Lakhs	Qty	Value (Rs.) in Lakhs	
1998	67236	9221	2651	954	10%
1999	60075	7860	1622	985	13%
2000	64563	14439	2284	2565	18%
2001	69386	15277	874	2210	15%
2002	67563	13318	2299	1577	12%
2003	83756	15530	6656	4700	30%
2004	84394	17196	5504	3909	22%
2005	103091	27110	9748	9346	34%
2006	96326	26382	9065	8564	32%
2007	91185	34446	8642	8660	25%
2008	88771	33269	7458	8013	24%
2009	80687	34635	10470	11133	32%
Maximum	103091	34635	10470	11133	34%
Minimum	60075	7860	874	954	10%
Mean	81583	21247.5	5672	6043.5	22%

Source: Directorate of Fisheries, Govt. of Goa Personally Computed (Personal Computation)

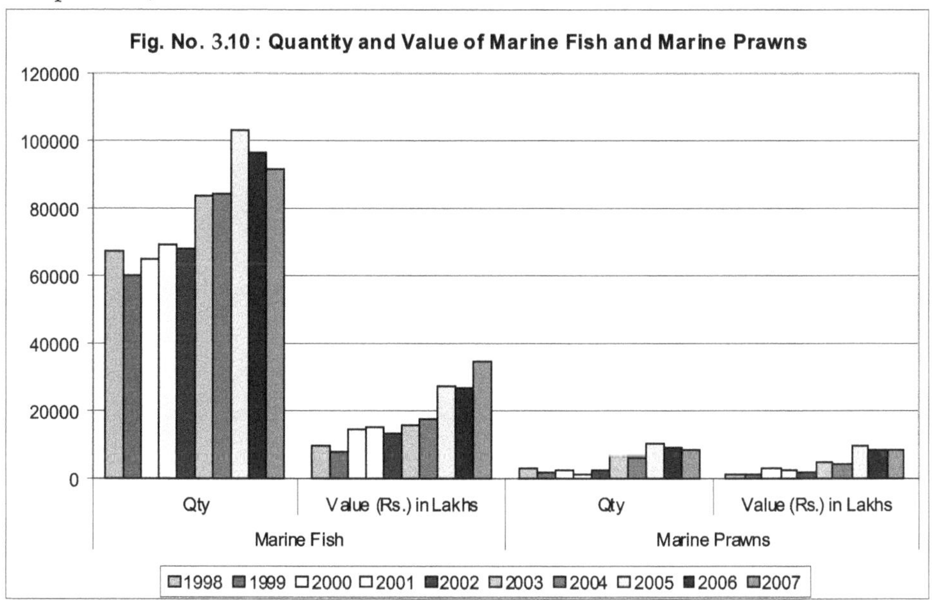

3.11 UTILIZATION OF MARINE FISH CATCH

The following table displays the trend of utilization of marine fish catch in the study area during the period 1971 to 2009-10. The quantity of fish i.e. marketed, sun dried, salted and miscellaneous varies from one decade to another and it appears that the quantity of fish catch was very low at 31502 metric tons in 1981-82 as compared to 1971-72, and then there was gradual increase in the fishery output, which reached the highest about 103091 metric tons of marine fish in 2005-06, out of which 92782 M.tons marketed, 5151 M.tons sub dried, 3090 M.tons salted and 2088 M.tons was used for miscellaneous purposes. Subsequently, the fishery output has been declined to 80687 tons during 2009-10, reflecting marginal decline in the utilization of fish in the study area.

Table-3.19 : Utilization of Marine Fish Catch During the Years 1971-2009 Quantity in Metric Tons

Sl. No.	Method of Utilization	1971 -72	1981 -82	1991 -92	2001 -02	2004 -05	2005 -06	2006 -07	2008 -09	2009 -10
1	Marketed Fish	27586	26890	34457	54771	75954	92782	88694	79893	72618
2	Sun Dried	3598	890	4430	7493	4220	5151	4816	4438	4034
3	Salted	2399	1314	2953	3506	2533	3090	2890	2663	2421
4	Miscellaneous	5997	2408	7384	3616	1687	2088	1926	1775	1616
	Total	39580	31502	49224	69386	84394	103091	96326	88771	81927

Source: Directorate of Fisheries, Govt. of Goa

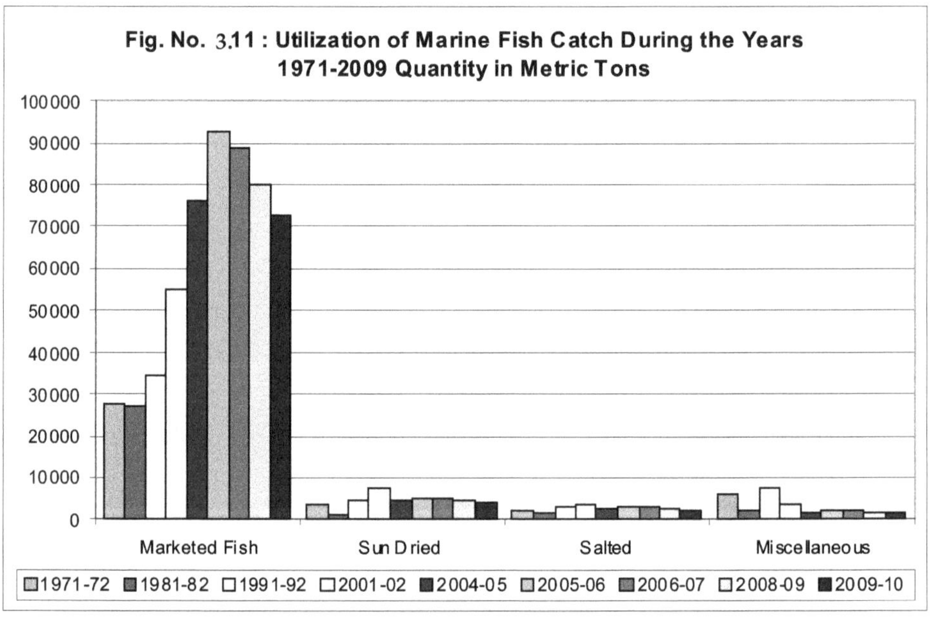

The study area has a total 36891 fishermen population, out of which 31727 are fulltime fishermen and 5164 others. Marine sector constitutes 15732 fishermen. Floating fishermen population is around 12000 whereas inland fishery sector has 9159 active fishermen population in Goa.

Table-3.20 : Marine Fishing – Active Fishermen Population

Marine Fish Landing Centres	Fishermen Population
Keri	70
Arambol	465
Morjim	543
Siolim	285
Chapora	259
Mandrem	185
Calangute	180
Candolin	84
Nerul	205
Cortabim	500
Velsao	189
Siridao	350
Nauxi	220
Causaubian	47
Issocian	78
Majorda	12
Khariwado	2798
Baina	1611
Bagmalo	288
Colva	622
Benaulim	1707
Cavellosim	404
Varca	310
Cutbona	911
Betul	1509
Saleri	245
Agonda	240
Palloleum	785
Nuuem	170
Talpona	166
Ghalsibag	140
Mashem	36
Pollem	316
Malim (Betim)	4500
Total	15732

Source: Directorate of Fisheries, Government of Goa (2007-08).

From the above table, it appears that different proportion of active fishermen population engaged in marine fishing activities from all over Goa. Malim Jetty records the highest number of active fishermen population, i.e., 4500 followed by Khariwado Vasco 2798 due to excessive fishing activities and large scale exploitation of Fishery Resource in the study area, whereas Majorda records less number of fishermen population with just 12 as it is small scale fishing area of Goa. Benaulim and Betul have moderate size of active fishermen population, i.e., former 1707 and latter 1509 engaged in marine fishing activities in the study area.

The Profile of Marine Fishery Resources

The following table depicts the scenario of marine fishery resource in Goa.

Table-3.21 : Profile of Marine Fishing Fishery Resources of Goa

Location Code	Name of the Village	Area in Sq.kms.	Total Population	Density	Rate of Literacy	Fisherman Engaged in Fishing Activity	% of Population	Length of Coast (in kms.)
1	Keri	40.5	2853	70.4	76.2	70	2.4	0
2	Arambol	96.6	5117	53	76.3	465	9	1
3	Siolim	122	10318	84.5	80.8	287	2.7	-
4	Chapora	BPFI	-	-	-	259	-	3
5	Mandrem	195.9	8022	41	75.35	189	2.7	3
6	Morjim	99.7	6390	64	76.2	543	8.4	2
7	Vagator	BP	-	-	-	NA	-	3
8	Aajuna	130	8624	66	78.5	NA	-	1
9	Calangute	117	15783	135	171.46	180	1.14	4
10	Candolim	70	8604	123	74.58	84	0.97	1
11	Singuerim	BP	-	-	-	-	-	3
12	Nerul	NA	4698	-	77.18	205	4.3	0
13	Malim	BPJ	-	-	-	4500	-	5
14	Donapaula	BP	-	-	-	-	-	5
15	Nauxi	BP	-	-	-	220	-	3
16	Siridao	77.2	2872	37.2	64.86	350	12.1	2
17	Aggassaim	BP	-	-	-	210	-	3
18	Cortalim	83.3	6970	84	70.2	500	7.17	4
19	Chiklaim	120.0	7604	63	80.2	NA	-	5
20	Khariwaddo	BPJ	-	-	-	2798	-	4
21	Baina – NS	NS	-	-	-	1611	-	-

Contd...

Table-3.21 Contd..

22	Bagmalo	BP	-	-	-	288	-	5
23	Velsao	16.2	1431	88	77.35	489	34.17	5
24	Cansaulim	19.1	2333	122	76.2	47	2.0	-
25	Issocim	27.3	547	20	82.2	78	14.2	-
26	Majorda	43.8	2968	68	76.4	12	0.4	4
27	Colva	28.7	3719	129	64.96	622	16.72	5
28	Benaurim	100.0	10158	101	73.7	1707	16.80	2
29	Varca	80.0	4865	60	76.7	310	6.3	-
30	Quelossim	106.4	2556	24	61.4	404	15.8	3
31	Cutbona	FJ	-	-	-	911	-	-
32	Betul	BPFJ	-	-	-	1509	-	4
33	Cabode Rama	BP	-	-	-	64	-	1
34	Saleri	NS	-	-	-	245	-	5
35	Agonda	148.0	3592	24	70.4	240	6.6	-
36	Palolem	BP	-	-	-	785	-	5
37	Nuvem	NS	-	-	-	170	-	-
38	Talpona	BP	-	-	-	166	-	4
39	Galgibag	NS	-	-	-	140	-	-
40	Mashem	NS	-	-	361	-	5	-
41	Polem	NS	-	-	316	5	-	-
42	Ambelim	-	-	-	71	-	-	-

Source: Directorate of Fisheries, Directorate of Census Operation (2010), Government of Goa.

GOA – POPULATION OF MARINE FISH LANDING CENTERS

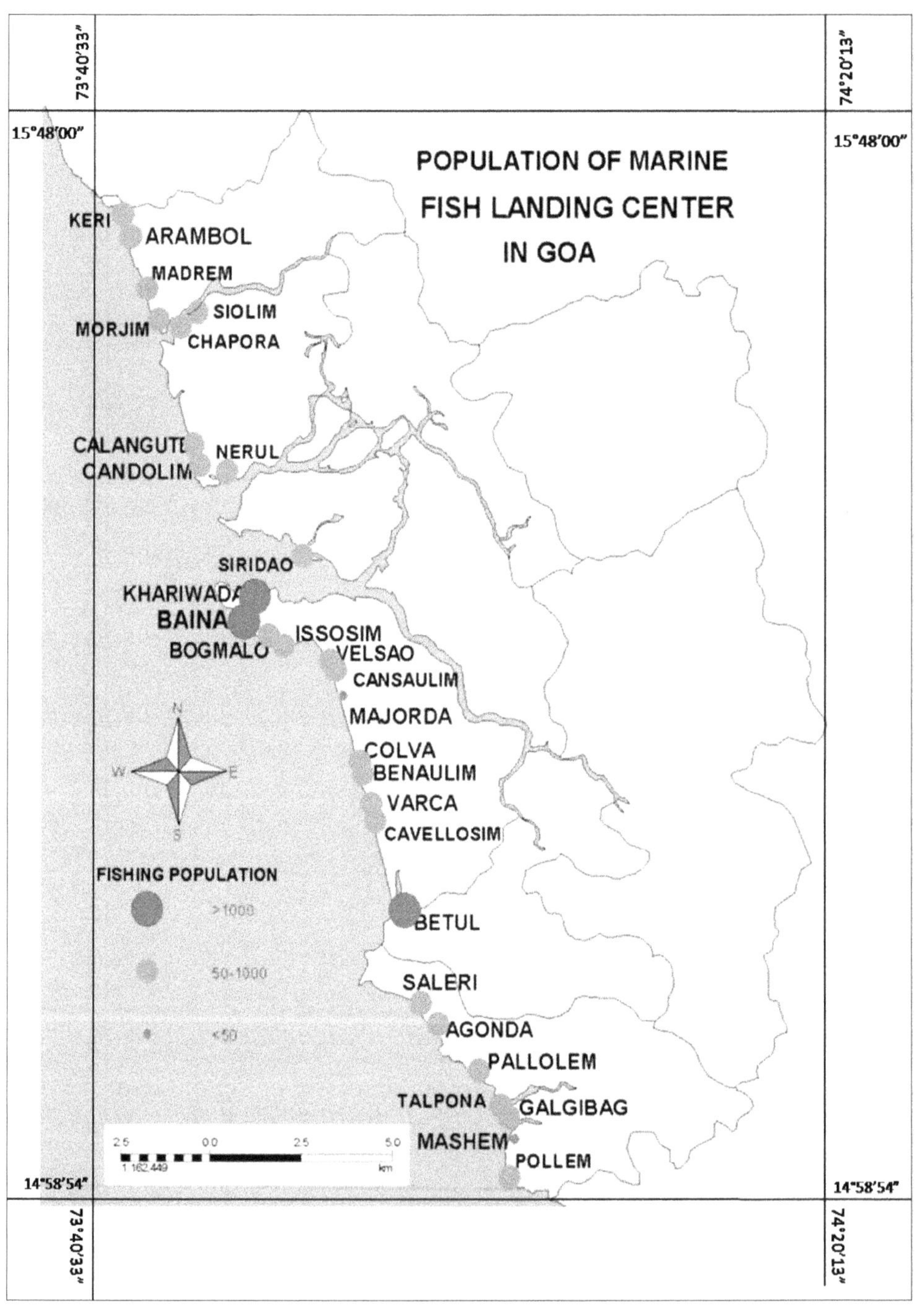

Map-3.2

GOA – COASTAL AND INLAND FISHING VILLAGES

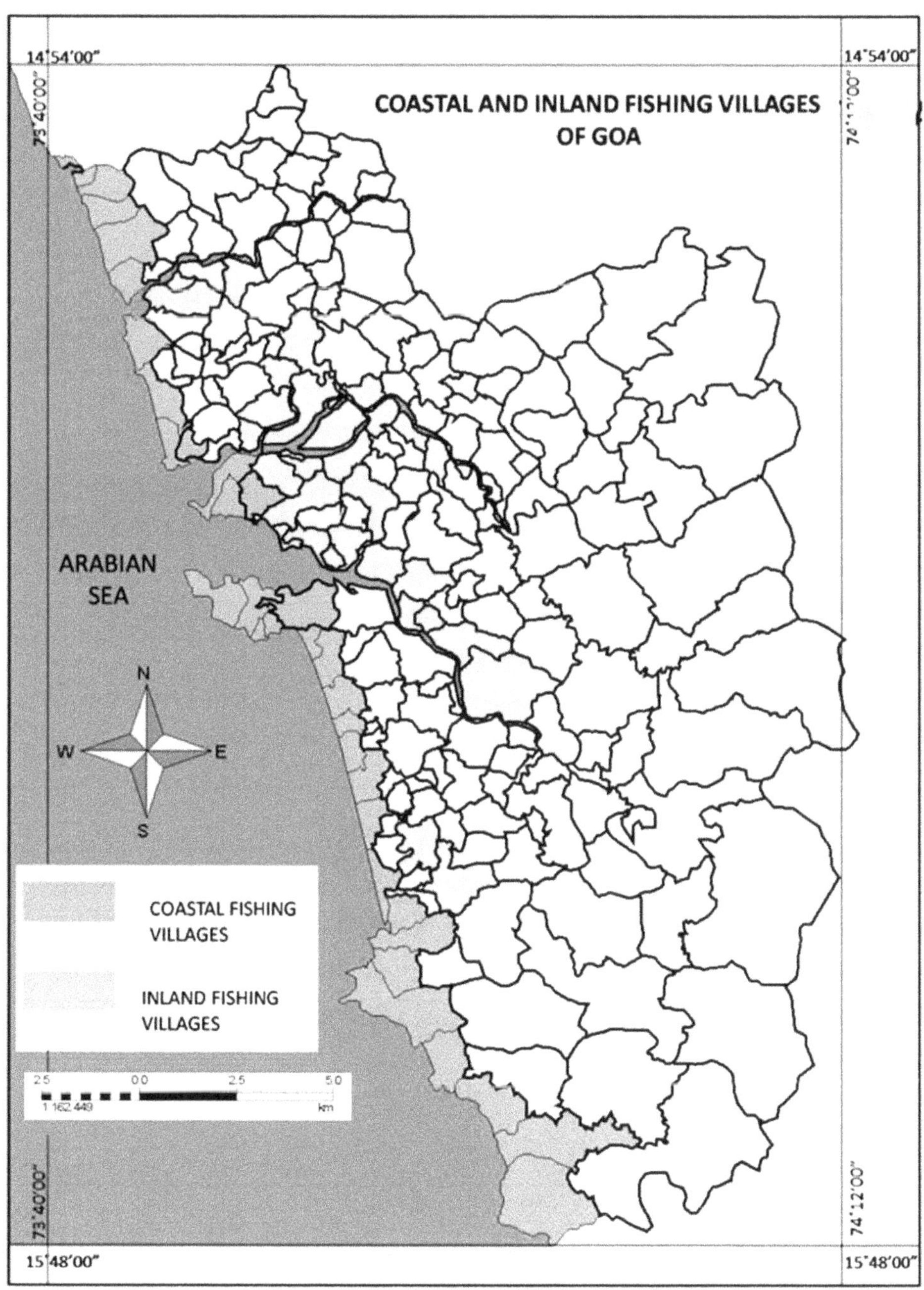

Map-3.3

From the table it reflects that different proportion of fishermen engages in fishing activities from different fishing villages in the study area. It has been observed that fishermen are quite literate and the average literacy has been above 70 percent in fishermen community. Each village or jetties have varied coastal length which is very useful for their fishing activities. Some villages do not have coast as they are located interior towards river bank side. Some of the villages i.e. Mashem, Galgibag, Polem, Ambelim are not settlements as per the 2001 census report, hence any records are not available on their size of population and geographical areas. Major fishing jetties witness large size of active fishermen participation in fishing activities compared to other areas.

As per the available data from the Directorate of Fisheries, Government of Goa (2010), there are 15732 active fulltime fishermen engaged in marine fishing activities from various marine fish landing centers in addition to 12000 floating fishermen population, which helps in successful fishing operations.

The following tables shows the responses of fishermen confined to their area of operation, method of fishing and the type of vessel they use has been observed with the help of primary survey by using random sampling technique by the researcher.

Table-3.22 : Fishing Operations

Area of Operations	Respondents	Percentage
Shallow water	55	55
Deep seas/shallow	27	27
Deep sea	18	18
Total	100	100

Source: Field Survey by the Researcher (2010).

The above table reveals that, as per the investigation, 55 percent of fishermen practice their fishing activities in shallow water, 27 percent in both deep sea as well as shallow waters to get big catch, whereas 18 percent go to deep sea for fishing.

Table-3.23 : Method of Fishing

Method	Respondents	Percentage
Traditional/by Rowing	46	46
Mechanized canoes	40	40
Trawler	14	14
Total	100	100

Source: Field Survey by the Researcher (2010).

The above table reveals that, 46 percent of fishermen are using traditional methods, i.e., by rowing boats, 40 percent are operations mechanized canoes and 14 percent are making use of trawler for their commercial fishing operation.

Table-3.24 : Type of Vessel they Use

Type in Length & Weight	Respondents	Percentage
22 feet, 3 tons	56	56
10 feet, 1.9 tons	36	36
21 feet, 3 tons	8	8
Total	100	100

Source: Field Survey by the Researcher (2010).

The above table reveals that, 56 percent of fishermen are fishing vessels measuring 22 feet length and weighing 3 tons. 36 percent use vessel of 10 feet length and weighing 1.9 tons, whereas 8 percent respondents said they are using boats of 21 feet length weighing 2 tons.

The fishermen supply their 80 percent of catch to near by major fish markets as well as minor centers and 20 percentage of fish goes to the hotels, bar and restaurants directly.

Some of the fishermen have financial as well as labour problems, hence they do not have any further expansion plans of their fishing activity.

The following table displays the distribution of mechanized and traditional boats at various marine fish landing centers of Goa during 2009-10.

Table-3.25 : Fish Landing Centers (Marine) with the Number of Mechanized/Non-Mechanized Boats

Sl. No.	Marine Centres	Mechanized Canoes	Non-Mechanized Canoes	Total
1	Sirido	35	20	55
2	Nauxi	16	18	34
3	Morjim	34	20	54
4	Siolim	23	32	55
5	Arambol	30	27	57
6	Mandrem	7	14	21
7	Calangute	98	20	118
8	Candolim	20	8	28
9	Nerul	85	10	95
10	Velsao	79	60	139
11	Cansaulim	45	40	85
12	Issosim	8	13	21
13	Majorda	5	9	14
14	Khariwada	0	42	42
15	Baina	63	52	115
16	Bogmalo	24	35	59
17	Colva	0	20	20
18	Benaulim	12	34	46
19	Cavellosim	7	20	27
20	Varca	9	26	35

Contd...

Table-3.25 Contd..

21	Saleri	32	9	41
22	Agonda	16	6	22
23	Pallolem/Colombo	55	15	70
24	Nuvem/Matvem	5	7	12
25	Talpona	8	10	18
26	Galzibag	3	5	8
27	Pollolem	9	2	11
28	Mashem	0	5	5
	Total	728	579	1307

Source: Directorate of Fisheries, Government of Goa (2010)

3.12 THE ROLE OF MECHANIZATION

In exploitation of fisheries, mechanization implies the use of advanced techniques in extraction of fishery resource, i.e., equipment, machine boats vessel, etc.

In the year 1957, the Portuguese Government introduced mechanization with modern boats. Under the circumstances, the mechanization did not work properly, and the activity incurred heavy losses. The project was not successful, and it was only after the liberation of Goa in the year 1961 that the situation changed. The region was under the control of Central Government of India, as it was a Union Territory, i.e., Goa, Daman and Diu, and the Separate Department of Fisheries was created in the year 1963. Then the mechanization was given tremendous encouragement, thereby mechanized fishing came to be practiced on an intensive scale.

Under the Department of Fisheries, Mechanization of fishing craft has been given a prominent place among other schemes for the development of fisheries. Engines are improved, types of mechanized boats were made available to the fishermen on loan cum subsidy basis. As a result, the numbers of mechanized boats, i.e., trawlers increased from 4 in 1961-62 to 110 in 1970 and then gradually increased to 186 in 1980, followed by 425 in 1993, 1092 in 2000, 1134 in 2003, 1152 in 2005 and 1157 in 2009-10 respectively over the years. There are 1157 mechanized trawlers registered with the Directorate of Fisheries as on 2005-06. After that registration of trawlers was stopped to prevent excessive mechanization in Goa's waters, which seemed to pose ecological threat to marine life, mangroves, etc.

During the last 30 years, the state has experienced high degree of mechanization in the fishing industry, and has led to greater fishery output, income, employment and growth of marketing activities.

During the last 5 year plan, the study area has developed rapid mechanization in fishing crafts, resulting in over 619 fishing canoes, 859 non-mechanized canoes operated in Goa's waters (258 of 619 mechanized canoes are engaged in inland fishing and around 728 country crafts without board engaged in gill net fishing). The mechanized trawlers concentrate on marine fish in the open sea. They usually operate in about 10 kms range at a depth of 20 meters.

GOA – MECHANIZED BOATS DISTRIBUTION

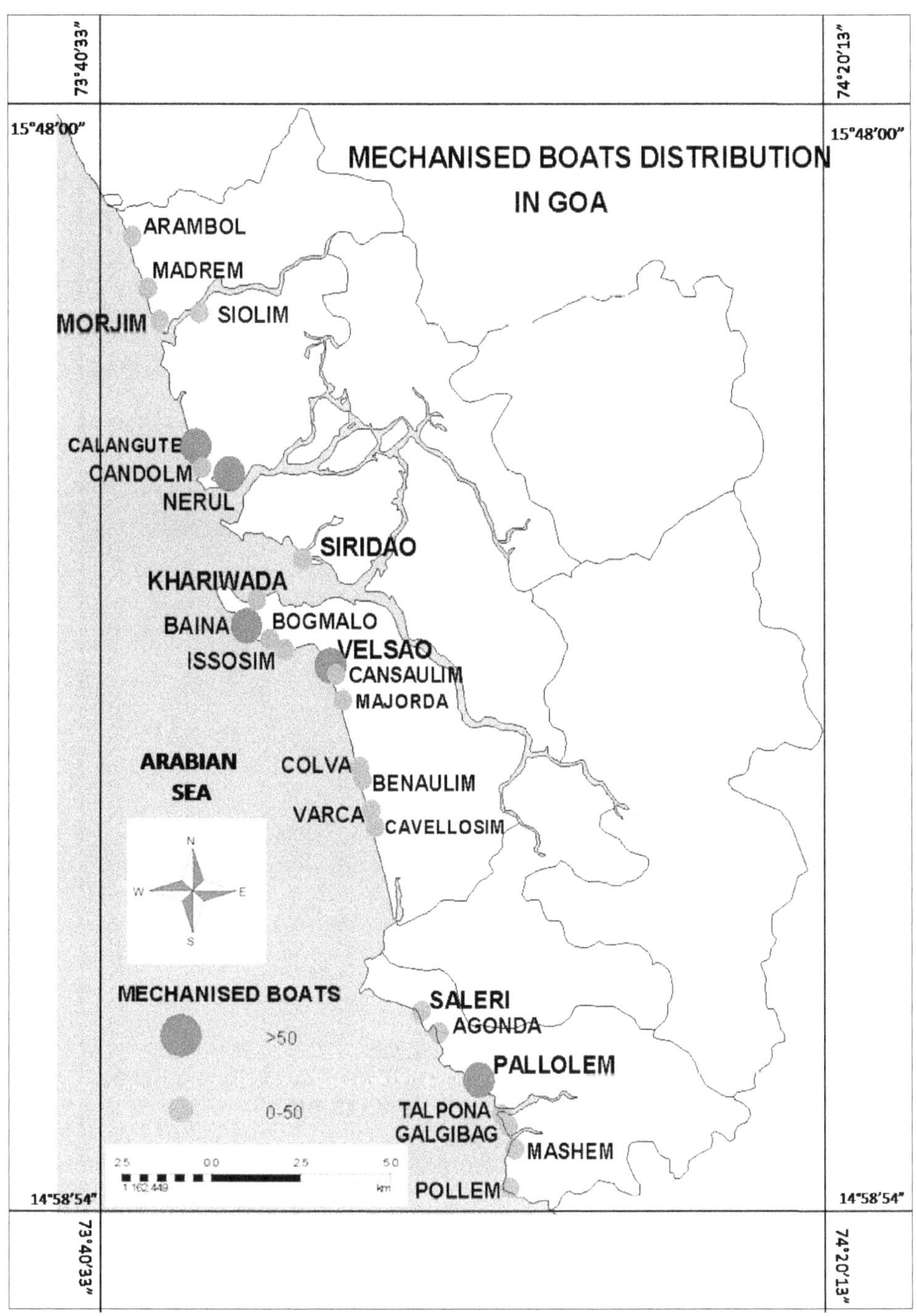

Map-3.4

GOA – NON-MECHANIZED BOATS DISTRIBUTION

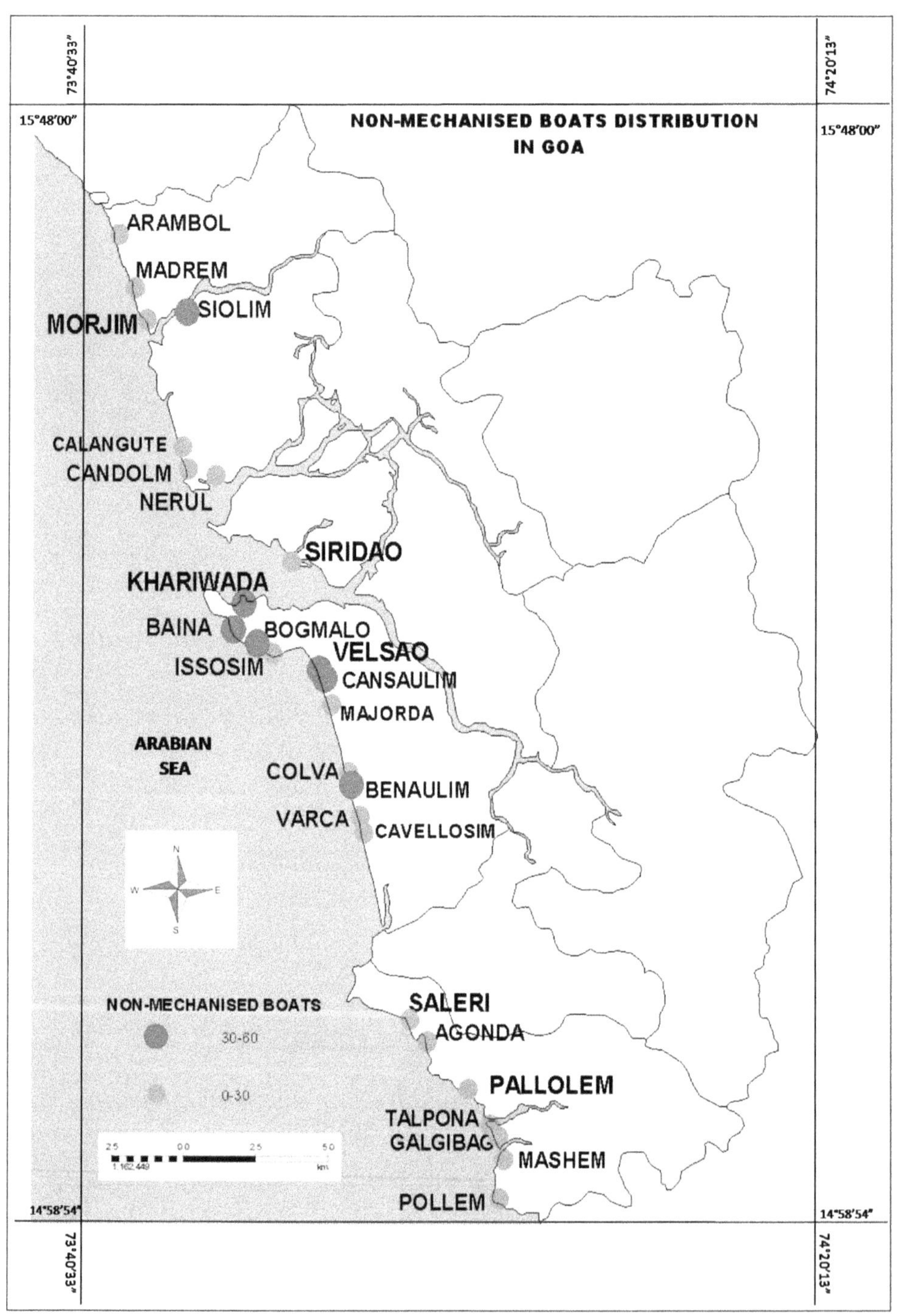

Map-3.5

The following table displays the trend of mechanized trawlers in fishery resource development, since the liberation of Goa from the Portuguese rule in 1961. The trend also exhibits the growth ratio of mechanized boats in the study area.

The role of mechanization in exploitation of fishery resources of Goa.

Table-3.26 : The Trend of Mechanized Trawlers in Goa's Waters

Year	No. of Trawlers	% of Growth
1961	4	-
1970	110	96
1980	186	40.8
1993	425	56.2
1994	473	10.1
1995	592	20.1
1996	617	4.05
1997	790	21.8
1998	912	13.3
1999	1056	13.6
2000	1092	3.2
2001	1128	3.1
2002	1134	0.5
2003	1134	0
2004	1152	1.5
2005	1157	0
2009	1157	0

Source: Directorate of Fisheries, Govt. of Goa (2009)

Table-3.27 : No. of Mechanized Trawlers in Fishing Operation at Major Fish Landing Centers of Goa as on 2008-09

Sl.No.	Fishing Centres	Fishing Trawlers in Operation	%
1	Chapora	75	6.4
2	Malim	350	30.2
3	Cutbona	270	23.3
4	Colva	80	6.9
5	Cortalim	30	2.5
6	Vasco	235	20.3
7	Talpona	50	4.3
8	Betul	30	2.5
9	Keri-Pernem	20	1.7
10	Other	17	1.4
	Total	1157	100

Source: Directorate of Fishers, 2007-08

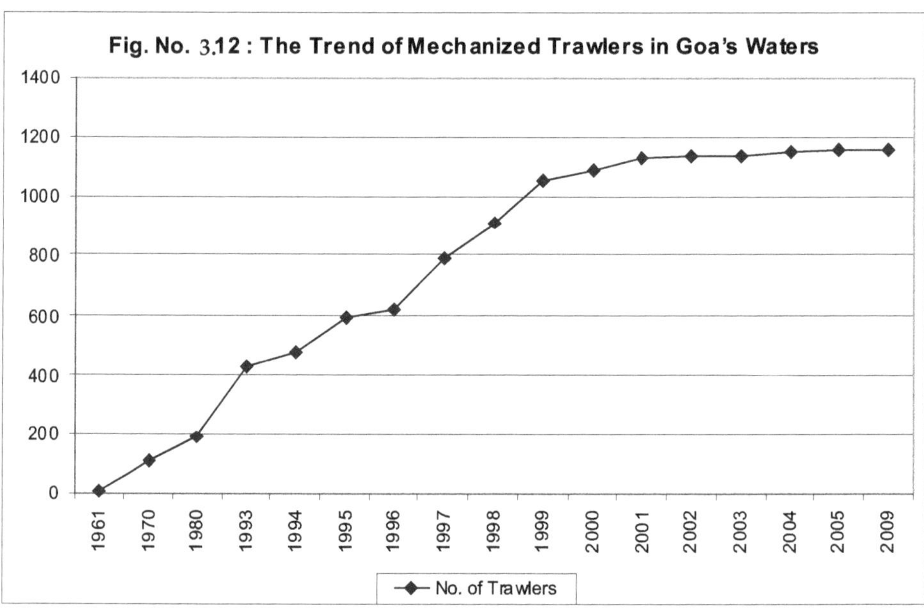

Fig. No. 3.12 : The Trend of Mechanized Trawlers in Goa's Waters

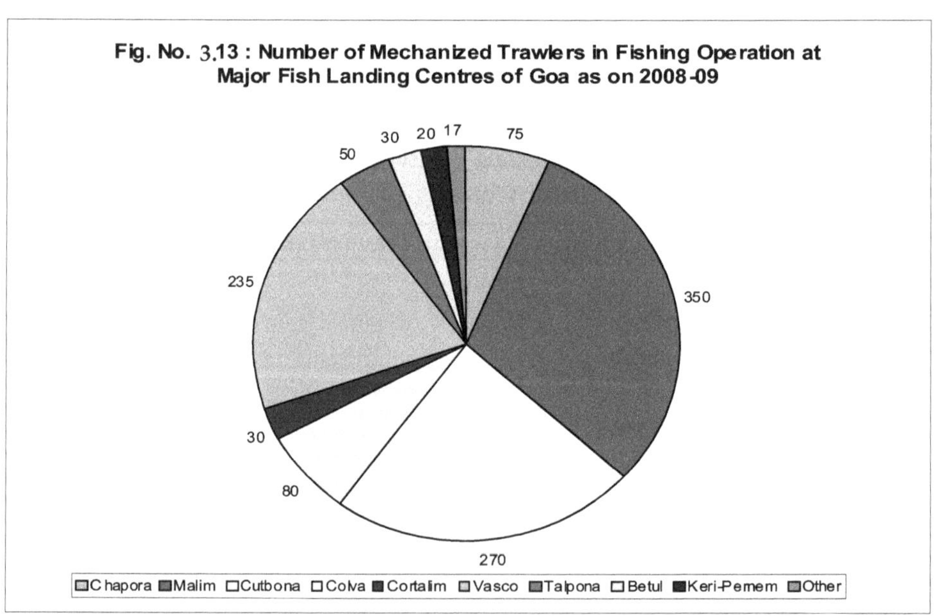

Fig. No. 3.13 : Number of Mechanized Trawlers in Fishing Operation at Major Fish Landing Centres of Goa as on 2008-09

Table-3.28 : Talukwise Distribution of Mechanized and Non-mechanized Canoes

Sl. No.	Taluk	Canoes Mechanized	Canoes Non-Mechanized	Total
1	Pernem	195	165	360
2	Bardez	126	127	253
3	Tiswadi	33	109	142
4	Salcete	140	242	382
5	Vasco	50	145	195
6	Canacona	71	61	132
7	Ponda	4	10	14
	Total	619	859	1478

Source: Directorate of Fishers, 2007-08

Growth of Fishing Fleet in Goa during 1960-61 to 2009-10

The following table displays the trend of growth of fishing fleet, which is the key factor in exploitation of fishery resources in the study area.

Table-3.29 : Growth of Fishing Fleet in Goa during 1960-61 to 2009-10

Sl. No.	Year	Mechanized Trawlers	Non-Mechanized Trawlers	Total
1	1961-62	4	4125	4129
2	1970-71	110	3600	3710
3	1980-81	189	2887	3076
4	1985-86	600	2450	3050
5	1990-91	707	2050	2757
6	1996-96	880	1980	2860
7	1997-98	1056	1840	2896
8	1998-99	1092	1890	2982
9	1999-2000	1092	2194	3286
10	2000-01	1128	1963	3091
11	2003-04	1134	1700	2834
12	2009-10	1157	859	2016

Source: Directorate of Fisheries, Government of Goa (2009-10).

From the above table it appears that the concept of mechanization favoured the growth of mechanized trawlers from just 4 in the year 1961-62 to 1157 in 2009-10. This has given tremendous encouragement to the growth of fishing industry in general and fishery resource in particular, whereas the number of non-mechanized canoes has come down drastically from 4125 in 1961-62 to 859 in the year 2009-10. This is mainly due to change in methods, application of new technology in exploitation of fishery resources by the fishermen community in the study area. The over fishing and overcrowded boats must be strictly prohibited. Otherwise the famous tragedy of common will soon be experienced (Koli, 2008).

Hence the hypothesis that the modern physical infrastructure has encouraged fishing activities rather than the traditional system has been tested with literature and confirmed the hypothesis.

Fig. No. 3.14 : Talukwise Distribution of Mechanized and Non-mechanized Canoes

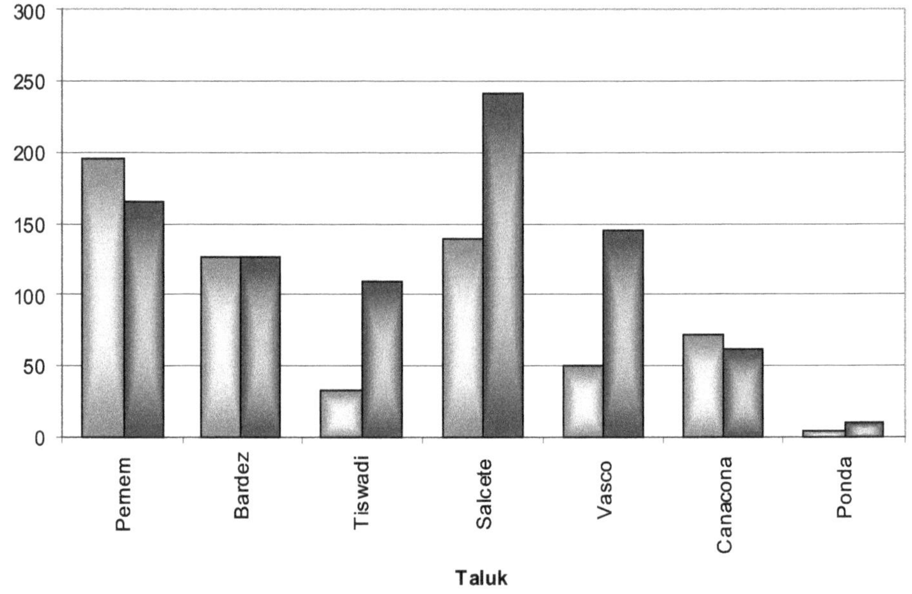

300
250
200
150
100
50
0

Pemem
Bardez
Tiswadi
Salcete
Vasco
Canacona
Ponda

Taluk

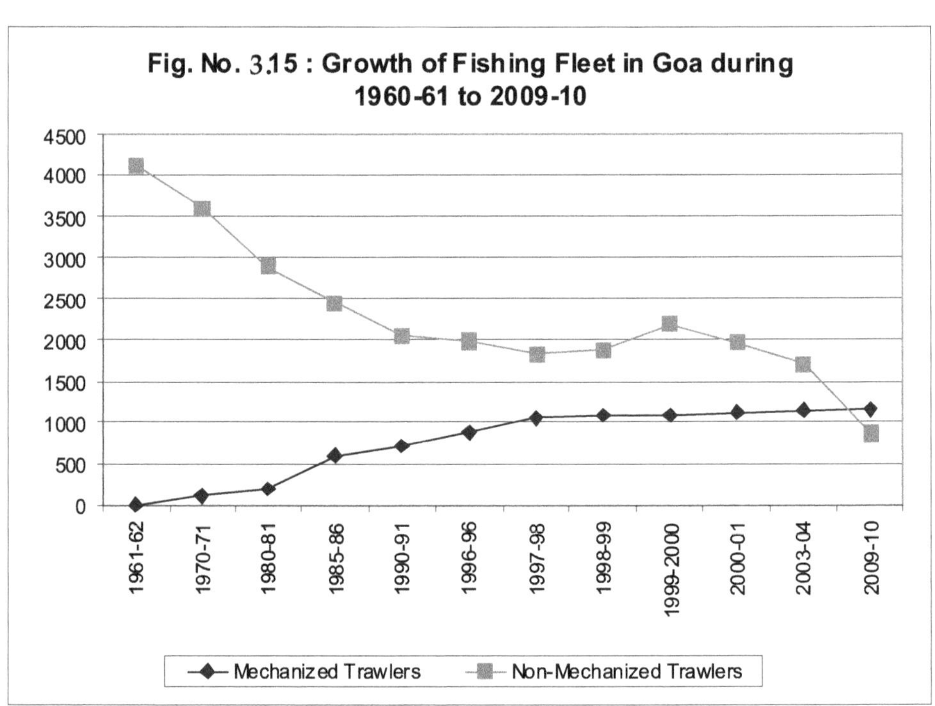

GOA – FISHING JETTIES WITH THE NUMBER OF TRAWLERS

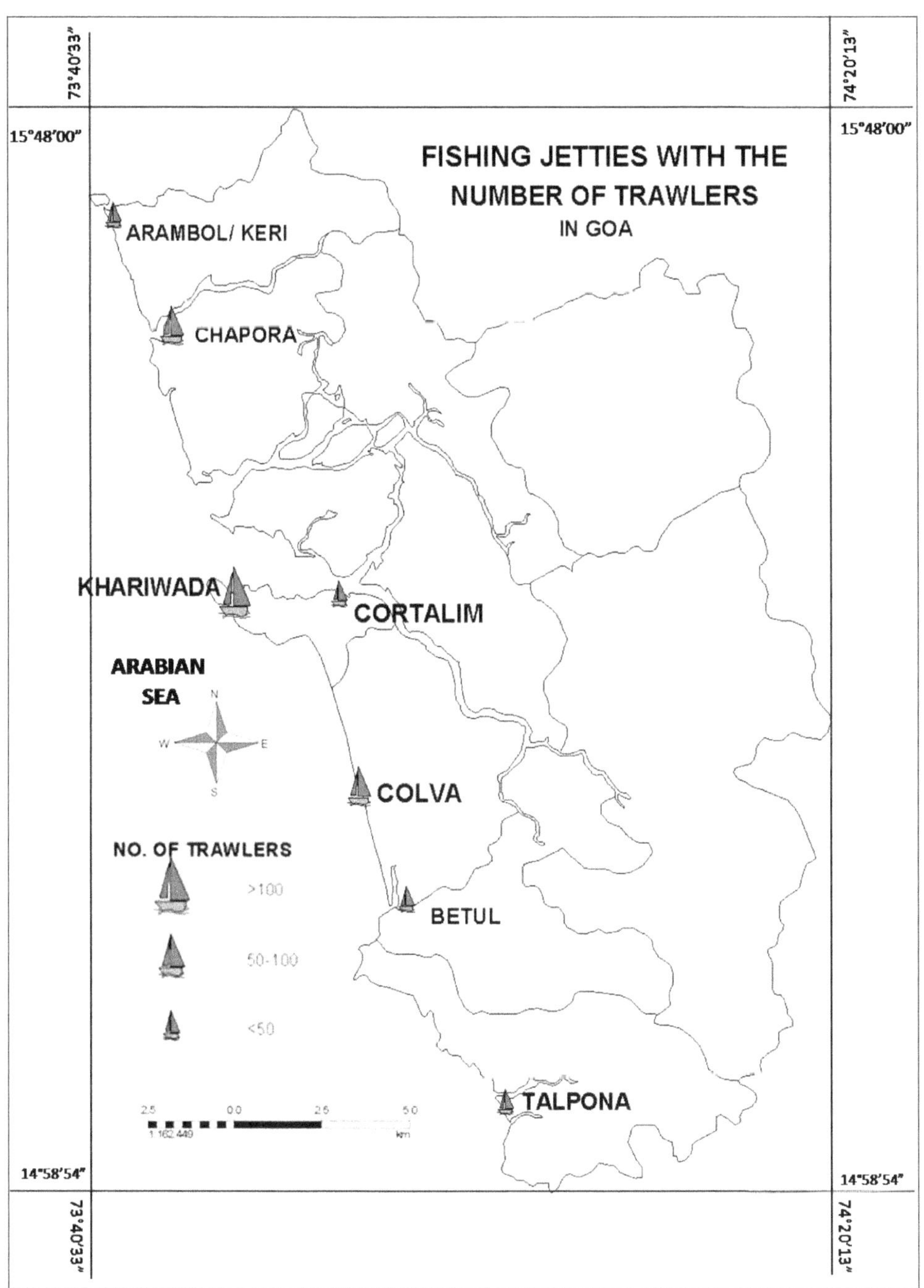

Map-3.6

3.13 MARINE FISHERY RESOURCE PRODUCTION AND OUTPUT VALUES

So far as fishery resource production and output values are concerned, normally the output values depend upon the fisheries production. Higher productivity, comparative records, a particular taluk is a leading producer of fisheries and output values in the state. The quantity of fish produced in Goa is about 80,687 metric tons, and the output value is Rs. 346.35 crores in the year 2009-10. This is followed by 91185, MT output value Rs. 344.46 crores during 2007-08, 103091 metric tons, output value Rs. 271.10 crores in 2005-06, 83756 metric tons and output value is Rs. 155.30 crores during 2003-04 and also comparatively lower fishery output recorded at 69386 and valued at Rs. 152.77 crores by 2001-02 respectively. These are the total amount of fish catch and output values in the state as compared to inland fish production in the study area, which is around 3283 metric tons, and the output value is Rs. 2046 (lakhs) in the year 2009-10, followed by 3070 metric tons values at 1856 (lakhs) during 2007-08 and 4194 metric tons amounts to Rs. 1989 lakhs in 2005-06. The inland fish catch is limited due to less demand and consumption in the study area as many local and outside people are more towards marine fish as it gives inclined good taste and great variety. The following tables - 1, 2, 3, 4 & 5 show the production and output values from 2009-10 to 2001-02 (2001- 2009-10).

During 2009-10 Marmugao taluk ranks first in the state with production potential of 39448 metric tons of marine fish annually, which is about 48.89 percent of total fisheries production, and is worth about Rs. 169.33 crores, which accounts for 48.89 percent of total output value in the state.

The taluk has numerous creeks, bays, rocky island and an important fish landing center Khariwaddo Vasco where 36.4 percent of active fishermen population is engaged in the fishery activities. Good infrastructure, banking, insurance and availability of labour, etc. is influencing to a great extent development of fishery resources. The fishery production and output values were i.e. 42383 metric tons V = Rs. 160.06 crore in 2007-08, 31065 metric tons V = Rs. 81.69 crores during 2005-06, Q=19775 metric tons values at Rs. 36.66 crores in 2003-04 and the taluk has recorded the lowest ever fishery output at 7375 metric tons and the output value reached a meager amount of Rs. 16.23 crores in the year 2001-02. The trend shows that the highest output was recorded in the year 2007-08, while the lowest was during 2001-02. It has been observed that the fishery production and the output values slowly picked up, and the growth has been clearly evident in the Mormugao taluk, which has become an important fishing taluk, and a leading producer of fish in the study area.

Table-3.30 : Marine Fishery Resource Production and Output Values 2009-10

Sl. No.	Taluk	Fish Catching Landing Centers		Population in Marine Fishing Activities		Boats Mechanized		Boats Traditional		Production in Metric Tons		Value in Rs. Crores	
		No.	%	No.	%	No.	%	No.	%	No.	%	No.	%
1	Pernem	3	9	1193	7.58	71	9.75	61	7.1	674	0.83	2.89	0.83
2	Bardez	5	15.5	1085	6.89	226	31.04	165	19.43	26309	32.6	112.9	32.6
3	Tiswadi	2	6	570	5.36	51	7	127	14.95	411	0.5	1.76	0.5
4	Marmugao	9	27.2	5727	36.4	231	31.73	242	28.5	39448	48.89	169.3	48.8
5	Salcete	6	18.18	5059	32.1	21	2.8	109	12.8	12528	15.5	53.77	15.5
6	Quepem	-	-	-	-	-	-	-	-	-	-	-	-
7	Canacona	8	24.2	2098	13.3	128	17.5	145	17	1317	1.6	5.65	1.6
	Total	**33**	**100**	**15732**	**100**	**728**	**100**	**849**	**100**	**80687**	**100**	**346.4**	**100**

Source: Directorate of Fisheries (2009-10), Government of Goa, Personal Computation.

Table-3.31 : Marine Fishery Resource Production and Output Values 2007-08

Sl. No.	Taluk	Fish Catching Landing Centers		Population in Marine Fishing Activities		Boats Mechanized		Boats Traditional		Production in Metric Tons		Value in Rs. Crores	
		No.	**%**	**No.**	**%**	**No.**	**%**	**No.**	**%**	**No.**	**%**	**No.**	**%**
1	Pernem	3	9	1193	7.58	71	11.5	61	7.1	8000	8.7	30.22	8.7
2	Bardez	5	15.5	1085	6.89	195	31.7	165	19.43	15982	17.5	60.37	17.5
3	Tiswadi	2	6	570	5.36	126	20.4	127	14.9	10083	11	38.08	11
4	Marmugao	9	27.2	5727	36.4	140	22.7	242	28.5	42383	46.4	160.1	46.4
5	Salcete	6	18.18	5059	32.1	33	5.3	109	12.8	14595	16	55.13	16
6	Quepem	-	-	-	-	-	-	-	-	-	-	-	-
7	Canacona	8	24.2	2098	13.3	50	8.1	145	17	153	0.16	0.57	0.16
	Total	**33**	**100**	**15732**	**100**	**615**	**100**	**849**	**100**	**91185**	**100**	**344.5**	**100**

Source: Directorate of Fisheries (2007-8), Government of Goa, Personal Computation.

Table-3.32 : Marine Fishery Resource Production and Output Values 2005-06

Sl. No.	Taluk	Fish Catching Landing Centers		Population in Marine Fishing Activities		Boats Mechanized		Boats Traditional		Production in Metric Tons		Value in Rs. Crores	
		No.	%	No.	%	No.	%	No.	%	No.	%	No.	%
1	Pernem	3	9.0	1193	7.58	71	11.5	61	7.1	13265	12.8	34.88	12.8
2	Bardez	5	15.5	1085	6.89	195	31.7	165	19.43	21999	21.3	57.85	21.3
3	Tiswadi	2	6.0	570	5.36	126	20.4	127	14.95	20317	19.7	53.43	19.7
4	Marmugao	9	27.2	5727	36.4	140	22.7	242	28.5	31065	30.1	81.69	30.1
5	Salcete	6	18.18	5059	32.1	33	5.3	109	12.8	16445	15.9	43.24	15.9
6	Quepem	-	-	-	-	-	-	-	-	-	-	-	-
7	Canacona	8	24.2	2098	13.3	50	8.1	145	17.0	418	.004	1.1	.004
	Total	**33**	**100**	**15732**	**100**	**615**	**100**	**849**	**100**	**103091**	**100**	**271.10**	**100**

Source: Directorate of Fisheries (2005-06), Government of Goa, Personal Computation.

Table-3.33 : Marine Fishery Resource Production and Output Values 2003-04

Sl. No.	Taluk	Fish Catching Landing Centers		Population in Marine Fishing Activities		Boats Mechanized		Boats Traditional		Production in Metric Tons		Value in Rs. Crores	
		No.	%	No.	%	No.	%	No.	%	No.	%	No.	%
1	Pernem	3	9.0	1193	7.58	112	11.5	122	7.1	9224	11.01	17.10	11.01
2	Bardez	5	15.5	1085	6.89	307	31.7	330	19.43	18521	22.11	34.34	22.11
3	Tiswadi	2	6.0	570	5.36	198	20.4	255	15	13836	16.51	25.65	16.51
4	Marmugao	9	27.2	5727	36.4	220	22.7	484	28.5	19775	23.61	36.66	23.61
5	Salcete	6	18.18	5059	32.1	52	5.3	219	12.8	21894	26.14	40.59	26.14
6	Quepem	-	-	-	-	-	-	-	-	-	-	-	-
7	Canacona	8	24.2	2098	13.3	79	8.13	290	17	505	0.60	0.93	0.60
	Total	**33**	**100**	**15732**	**100**	**968**	**100**	**1700**	**100**	**83756**	**100**	**155.30**	**100**

Source: Directorate of Fisheries (2003-04), Government of Goa, Personal Computation.

Table-3.34 : Marine Fishery Resource Production and Output Values 2001-02

Sl. No.	Taluk	Fish Catching Landing Centers		Population in Marine Fishing Activities		Boats Mechanized		Boats Traditional		Production in Metric Tons		Value in Rs. Crores	
		No.	**%**	**No.**	**%**	**No.**	**%**	**No.**	**%**	**No.**	**%**	**No.**	**%**
1	Pernem	3	9.0	1193	7.58	124	11.5	134	7.1	8182	11.79	18.01	11.79
2	Bardez	5	15.5	1085	6.89	339	31.7	364	19.43	25173	36.27	55.42	36.27
3	Tiswadi	2	6.0	570	5.36	218	20.4	281	14.95	12274	17.68	27.02	27.02
4	Marmugao	9	27.2	5727	36.4	243	22.7	532	28.5	7375	10.62	16.23	16.23
5	Salcete	6	18.18	5059	32.1	57	5.3	241	12.8	15543	22.40	34.22	34.2
6	Quepem	-	-	-	-	-	-	-	-	-	-	-	-
7	Canacona	8	24.2	2098	13.3	88	8.13	319	17	839	1.2	1.84	1.2
	Total	**33**	**100**	**15732**	**100**	**1069**	**100**	**1866**	**100**	**69386**	**100**	**152.77**	**100**

Source: Directorate of Fisheries (2001-02), Government of Goa, Personal Computation.

Apart from Marmugao taluk, the Bardez taluk has emerged as second leading producer of fish and fish products with production potential of 26309 metric tons, which accounts for 32.6 percent of total output in the year 2009-10, which is valued at Rs. 112.93 crores, followed by Q = 15982 value 60.37 crores during 2007-08, Q = 21999 metric tons V = 57.85 crores in 2005-06, Q = 18521 metric tons V = 34.34 crores and the fishery production and output values were around 25173 metric tons valued at Rs. 55.42 crores during 2001-02. Over a period of ten years, the least output was recorded in the year 2007-08, whereas the highest production and output values were seen during 2009-10. Thus, it appears that the fishery production has been stagnated between the range of 15000 metric tons and 25000 metric tons.

Salcete taluk ranks third in the fishery resource production, and so far as output values are concerned, the taluk has produced about 12528 metric tons of fish, which is valued at Rs. 53.77 crores in the year 2009-10, which accounts for 15.5 percent of total fishery output and values in the study area. The taluk has been progressing very well and consistently in the past ten years, producing a good quantity of fish, which is widely marketed in southern parts of Goa. The taluk has produced about Q = 14595 metric tons V = Rs. 55.13 crores in 2007-08, Q = 16445 metric tons V = Rs. 43.24 crores during 2005-06, Q = 21894 metric tons V = Rs. 40.59 crores by 2003-04 and the quantity and output values were 15543 metric tons valued at Rs. 34.22 crores during 2001-02.

Canacona taluk is also becoming an important producer of fish and fishery products. Over the years, the progress is clearly evident as the fishery output and values are increasing gradually in the study area with 1317 metric tons valued at Rs. 5.65 crores in the year 2009-10, followed by Q = 153 metric tons V = Rs. 0.57 crore during 2007-08, Q = 418 metric tons V = Rs. 1.1 crore by 2005-06, Q = 505 metric tons V = Rs. 0.93 crore and the fishery output and values were recorded at 839 metric tons valued at Rs. 1.84 crores in the year 2001-02 respectively. The least output was observed in 2007-08 and the highest has been during 2009-10.

Pernem taluk is the northern most taluk, and has produced about 674 metric tons of fish, which is valued at Rs. 2.89 crores, which account for 0.82 percent of total output, and values are concerned in the study area in the year 2009-10, followed by 8000 metric tons V = 30.22 crores during 2007-08, Q = 13265 metric tons V = 34.88 crores in 2005-06, Q = 9224 metric tons V = Rs. 17.10 crores by 2003-04 and the fishery output was recorded at 8182 metric tons values at Rs. 18.01 crores during 2001-02 respectively.

Tiswadi taluk has produced about 411 metric tons of fish valued at Rs. 1.76 crores during 2009-10, followed by 10083 metric tons V = Rs. 38.08 crores in 2007-08, whereas the output was around 20317 metric tons valued at Rs. 53.43 crores by 2005-06, Q = 13836 metric tons V = Rs. 25.65 crores in 2003-04 and Q = 12274 metric tons of fish valued at Rs. 27.02 crores during 2001-02 respectively. The output has come down drastically inspite of a large number of mechanized, non-mechanized boats helping fishing activities in this taluk.

Quepem taluk is located along the coast of Goa. It has very short coastline, that is rocky approximately a kilometer. There are no records of fishing activities taking place in this area. In true sense, it is not a fishery taluk because the land and water inter-charge is very limited, as it is a rocky area, i.e., caboderama rocky headland affected fishing activities adversely in this region.

3.14 MARINE FISHERY RESOURCE DEVELOPMENT

The main objectives of the present study is to assess the fishery resource development for the period from 2001-02 to 2009-10 based on the various attributes that have already been discussed for specific point of time. The fishery resource development profile is calculated for the year (2001-02 to 2009-10) as revealed in the tables. The various attributes that have been selected for the index are:

1) Percentage share of marine fish catching centres in each taluk to the state total.
2) Percentage share of population involved in marine fishing activities in each taluk to the state total.
3) Percentage share of mechanized boats used in fishing activities in each taluk to the state total.
4) Percentage share of traditional boats operated for fishing in each taluk to the state total.
5) Percentage share of production of marine fisheries in each taluk to the state total.
6) Percentage share of value of fisheries caught in each taluk to the state total.

The above mentioned six attributes have been used to assess the level of fishery resource development, based on the index of ranking co-efficient given by Kendal (1968). The fishery resource development profile is calculated for the period from 2001-02 to 2009-10 as revealed in the tables that follow.

The following tables show composite index for fishery resources development of Goa with the help of personal computation.

Table-3.35 : Marine Fishery Resource Potentiality 2009-10

Sl. No.	Taluk	Fish Catching Landing Centers		Population in Marine Activities		Boats Mechanized		Boats Traditional		Production in Metric Tons		Value in Rs. Crores		COI
		%	Position	%	Position	%	Position	%	Position	%	Position	%	Position	
1	Pernem	9	5	7.58	4	9.75	4	7.1	6	0.83	5	0.83	5	4.83
2	Bardez	15.5	4	6.89	5	31.04	2	19.43	2	32.6	2	32.6	2	2.83
3	Tiswadi	6	6	5.36	6	7	5	14.95	4	0.5	6	0.5	6	5.5
4	Marmugao	27.2	1	36.4	1	31.73	1	28.5	1	48.89	1	48.89	1	1
5	Salcete	18.18	3	32.1	2	2.8	6	12.8	5	15.5	3	15.5	3	3.66
6	Quepem	-	-	-	-	-	-	-	-	-	-	-	-	-
7	Canacona	24.2	2	13.3	3	17.5	3	17	3	1.6	4	1.6	4	3.16

Source: Directorate of Fisheries (2009-10), Government of Goa, Personal Computation.

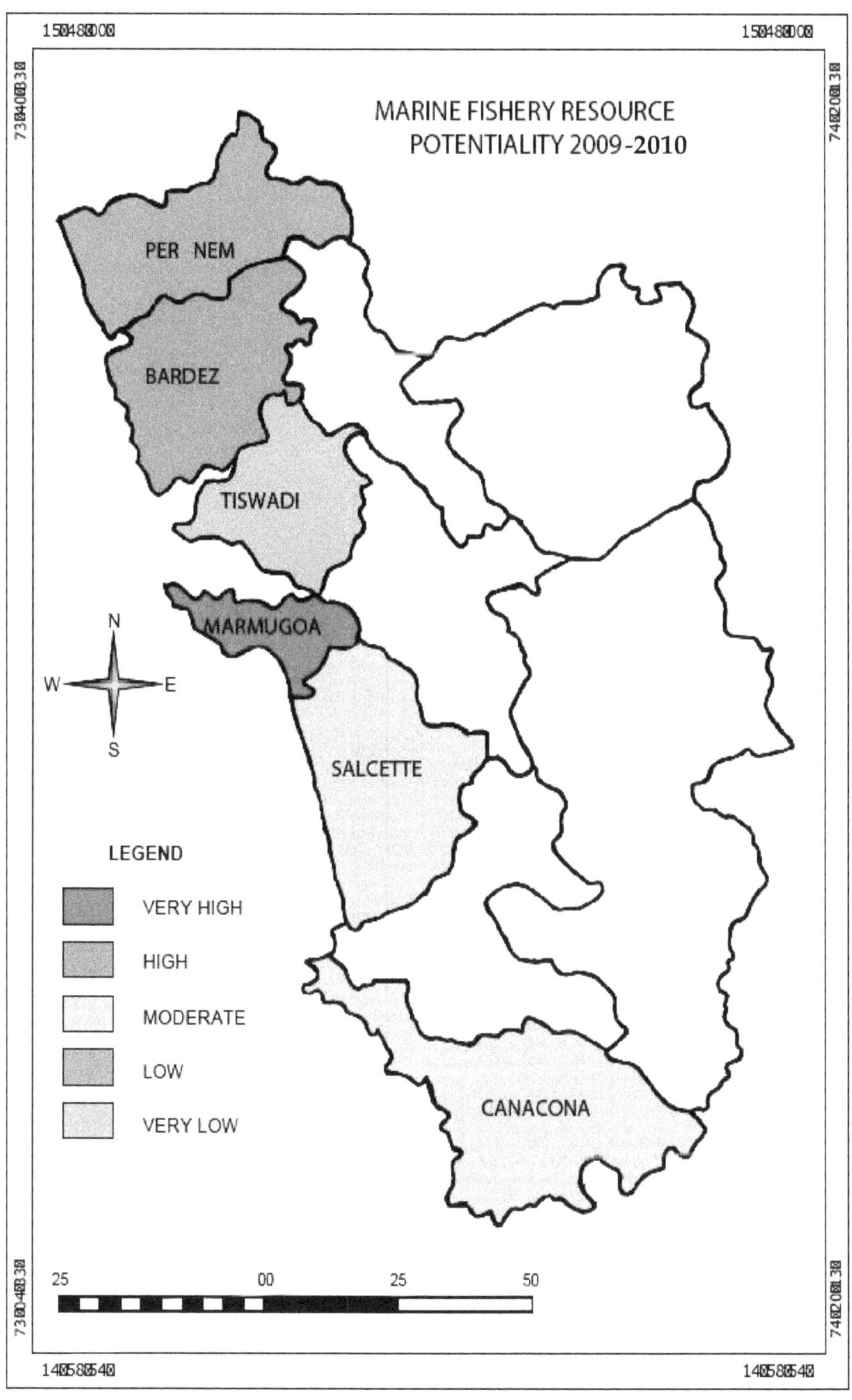

MARINE FISHERY RESOURCE
POTENTIALITY 2009-2010
PER NEM
BARDEZ
TISWADI
MARMUGOA
SALCETTE
CANACONA
N
W
E
S
LEGEND
VERY HIGH
HIGH
MODERATE
LOW
VERY LOW
25
00
25
50

Map-3.7

Table-3.36 : Marine Fishery Resource Potentiality 2007-08

Sl. No.	Taluk	Fish Catching Landing Centers		Population in Marine Activities		Boats Mechanized		Boats Traditional		Production in Metric Tons		Value in Rs. Crores		COI
		%	Position	%	Position	%	Position	%	Position	%	Position	%	Position	
1	Pernem	9.0	5	7.58	4	11.5	4	7.1	6	8.77	5	8.77	5	4.83
2	Bardez	15.5	4	6.89	5	31.7	1	19.43	2	17.52	2	17.52	2	2.6
3	Tiswadi	6.0	6	5.36	6	20.4	3	14.95	4	11.05	4	11.05	4	4.35
4	Marmugao	27.2	1	36.4	1	22.7	2	28.5	1	46.46	1	46.46	1	1.16
5	Salcete	18.1	3	32.1	2	5.3	6	12.8	5	16	3	16	3	3.6
6	Quepem	-	-	-	-	-	-	-	-	-	-	-	-	-
7	Canacona	24.2	2	13.3	3	8.1	5	17	3	0.16	6	0.16	6	4.16

Source: Directorate of Fisheries (2007-08), Government of Goa, Personal Computation.

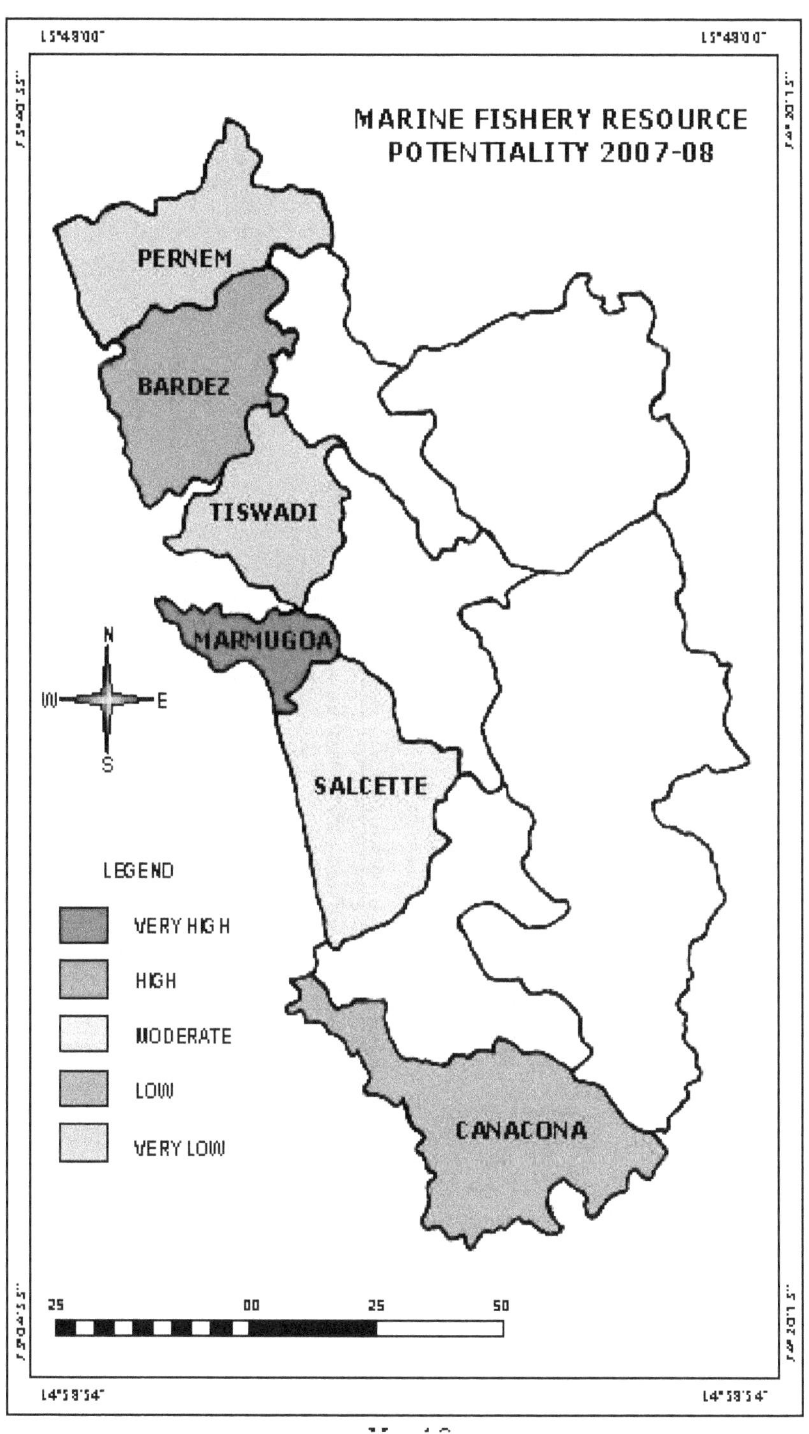

Map-3.8

Table-3.37 : Marine Fishery Resource Potentiality 2005-06

Sl. No.	Taluk	Fish Catching Landing Centers		Population in Marine Activities		Boats Mechanized		Boats Traditional		Production in Metric Tons		Value in Rs. Crores		COI
		%	Position	%	Position	%	Position	%	Position	%	Position	%	Position	
1	Pernem	9.0	5	7.58	4	11.5	4	7.1	6	12.8	5	12.8	5	4.83
2	Bardez	15.5	4	6.89	5	31.7	1	19.43	2	21.3	2	21.3	2	2.66
3	Tiswadi	6.0	6	5.36	6	20.4	3	14.95	4	19.7	3	19.7	3	4.16
4	Marmugao	27.2	1	36.4	1	22.7	2	28.5	1	30.1	1	30.1	1	1.16
5	Salcete	18.1	3	32.1	2	5.3	6	12.8	5	15.9	4	15.9	4	4.0
6	Quepem	-	-	-	-	-	-	-	-	-	-	-	-	-
7	Canacona	24.2	2	13.3	3	8.13	5	17	3	.004	6	.004	6	4.16

Source: Directorate of Fisheries (2005-06), Government of Goa, Personal Computation.

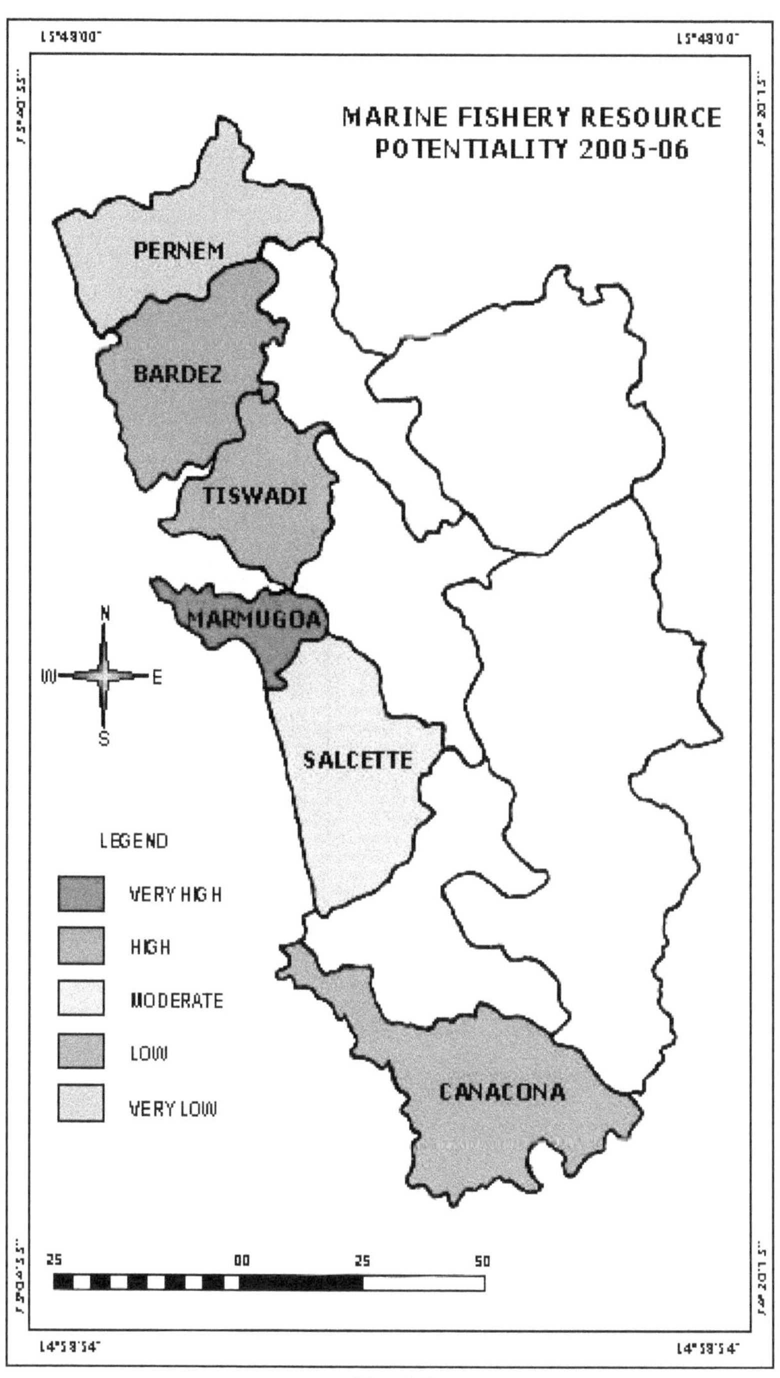

Map-3.9

Table-3.38 : Marine Fishery Resource Potentiality 2003-04

Sl. No.	Taluk	Fish Catching Landing Centers		Population in Marine Activities		Boats Mechanized		Boats Traditional		Production in Metric Tons		Value in Rs. Crores		COI
		%	Position	%	Position	%	Position	%	Position	%	Position	%	Position	
1	Pernem	9.0	5	7.58	4	11.5	4	7.1	6	11.01	5	11.01	5	4.83
2	Bardez	15.5	4	6.89	5	31.7	1	19.43	2	22.11	3	22.11	3	3.0
3	Tiswadi	6.0	6	5.36	6	20.4	3	15	4	16.51	4	16.51	4	4.5
4	Marmugao	27.2	1	36.4	1	22.7	2	28.5	1	23.61	2	23.61	2	1.5
5	Salcete	18.18	3	32.1	2	5.3	6	12.8	5	26.14	1	26.14	1	3.0
6	Quepem	-	-	-	-	-	-	-	-	-	-	-	-	-
7	Canacona	24.2	2	13.3	3	8.13	5	17	3	0.60	6	0.60	6	4.16

Source: Directorate of Fisheries (2003-04), Government of Goa, Personal Computation.

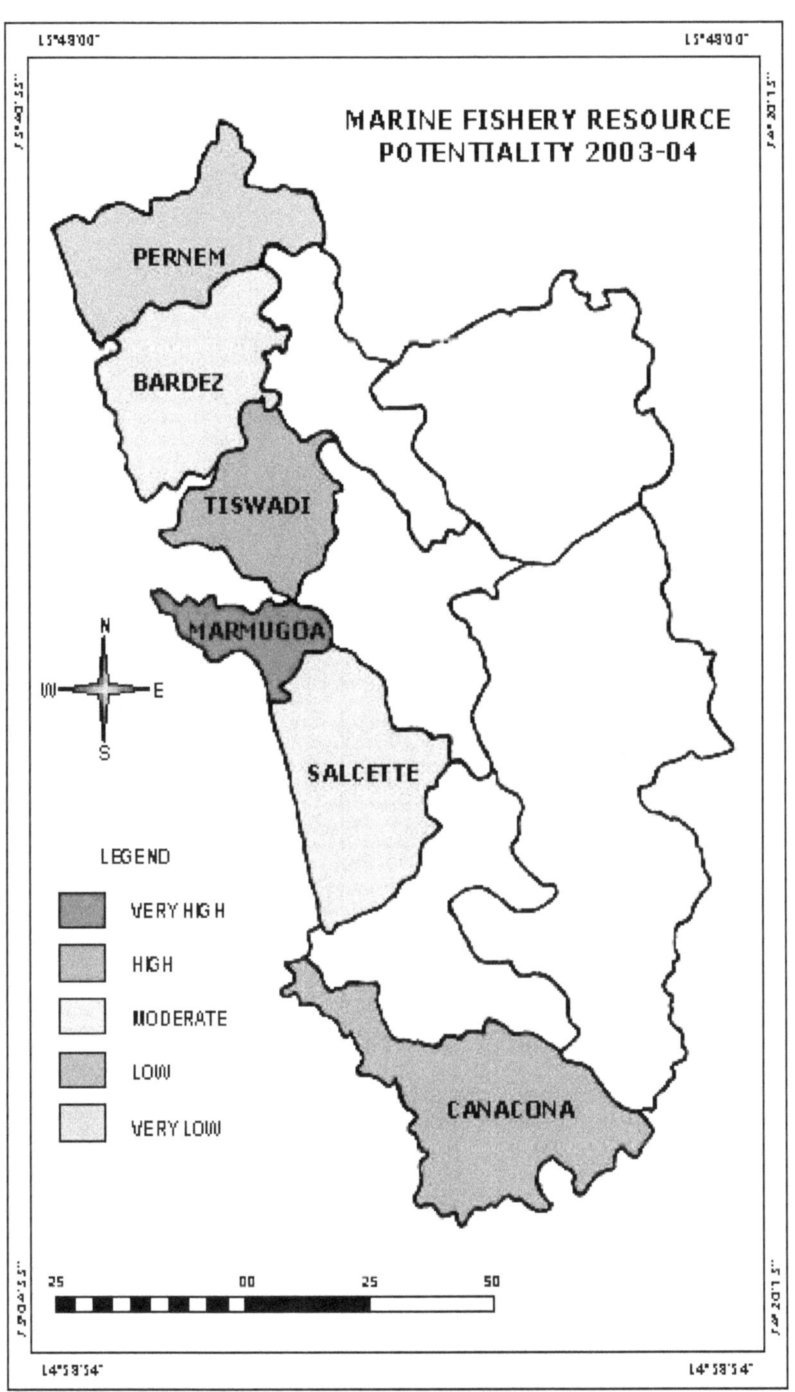

Map-3.10

Table-3.39 : Marine Fishery Resource Potentiality 2001-02

Sl. No.	Taluk	Fish Catching Landing Centers		Population in Marine Activities		Boats Mechanized		Boats Traditional		Production in Metric Tons		Value in Rs. Crores		COI
		%	Position	%	Position	%	Position	%	Position	%	Position	%	Position	
1	Pernem	9.0	5	7.58	4	11.5	4	7.1	6	11.79	4	11.79	4	4.5
2	Bardez	15.5	4	6.89	5	31.7	1	19.43	2	36.27	1	36.27	1	2.33
3	Tiswadi	6.0	6	5.36	6	20.4	3	14.95	4	17.68	3	17.68	3	4.16
4	Marmugao	27.2	1	36.4	1	22.7	2	28.5	1	10.62	5	10.62	5	2.5
5	Salcete	18.18	3	32.1	2	5.3	6	12.8	5	22.40	2	22.40	2	3.33
6	Quepem	-	-	-	-	-	-	-	-	-	-	-	-	-
7	Canacona	24.2	2	13.3	3	8.13	5	17	3	12	6	12	6	4.16

Source: Directorate of Fisheries (2001-02), Government of Goa, Personal Computation.

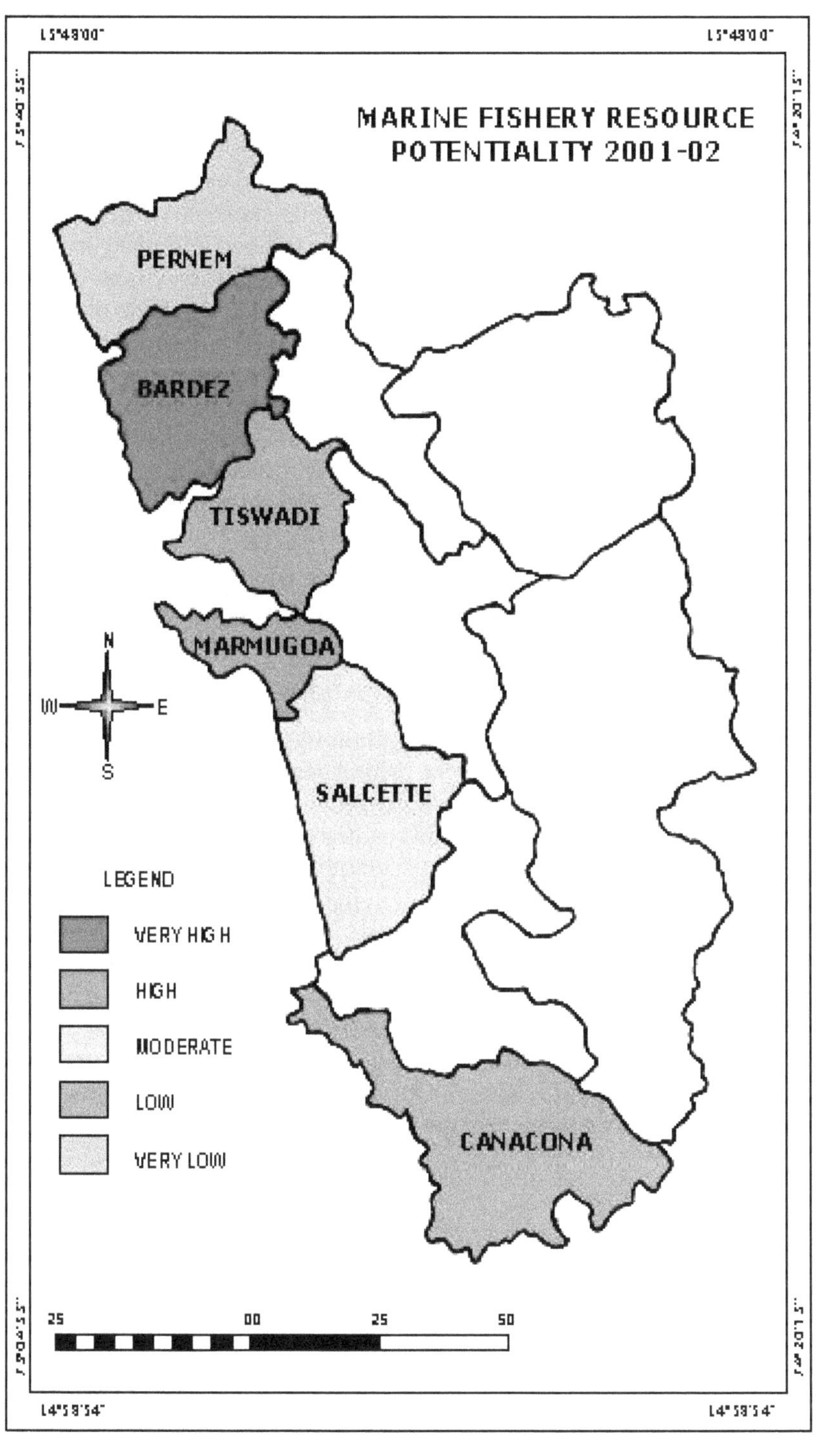

MARINE FISHERY RESOURCE
POTENTIALITY 2001-02
PERNEM
BARDEZ
TISWADI
MARMUGOA
SALCETTE
CANACONA
N
W
E
S
LEGEND
VERY HIGH
HIGH
MODERATE
LOW
VERY LOW
25
00
25
50

Map-3.11

Zone-1

The first zone comprises very high fishery resource development in the study area. It consists of one taluk Mormugao. This is the core area, which indicates the maximum development of marine fishery resources. This zone is characterized by broken coast with numerous bay and headlands, creeks, swamps, coral reefs, etc., situated well in the extreme west. Mormugao taluk has the largest number of fishermen population, which accounts for 36.4 percent of total fishermen in the state, engaged in fishing activities enthusiastically to earn their livelihood and make larger profits. This taluk has produced about 39448 tons of fish and fish products valued at Rs. 169.33 crores in the year 2009-10. This zone also indicates the use of mechanized boats and traditional boats for fishing operations, which account for nearly 22.7 percent and 28.5 percent respectively. Excellent infrastructure facilities, banking, insurance, labour, ready markets are encouraging fishery activities and fishery resource development in this region. Therefore, this zone can be regarded as a hub of fisheries in the study area. Further, this activity stimulates the growth of ancillary and subsidiary economic activities to provide employment opportunities to the people. Thus this zone can become highly prosperous in the near future.

Marmugao taluk has been leading producer which fall under the very high development zone from i.e. 2003-04 to 2009-10. But during 2001-02, Bardez taluk has recorded highest productivity of fisheries and output values in the study area.

Zone-2

Bardez taluka, which enjoys suitable conditions for fishery resource development is the second leading producer of marine fisheries for the last six years i.e. 2005-06 to 2009-10. This taluk has been placed in high development zone. Bardez taluka is a coastal taluk with 9 percent of fish catching centres and 7.8 percent of total fishermen population, engaged in fishing activities. It enjoys second largest position in fishery resource development. It enjoys first position in mechanized boats, second position in traditional boats, production, output values and fourth position in fish catching centres and it enjoys fifth position in fisherman population as well. However, this taluk has great potential for fishery resource development. With appropriate measures, fishery resource can be developed in this area on a large scale, thereby this zone could become highly progressive ensuring greater economic stability of fishermen community and of other people as well. During 2001-02, Mormugao taluk was placed in high development zone with its respective fishery output and values, etc.

Zone-3

Salcete taluk is the only taluk which falls in the moderate fishery resource development zone for the last 10 years. It has 3rd place in fish catching centre production and output values, second position in fishermen population, fifth position in traditional boats and 6th position in the use of mechanized boats. Salcete taluk is rich in labor for fishing activities, and favorable coast and other facilities are encouraging the development of fishery activities in this zone. During 2009-10, Canacona taluk has entered the moderate zone in respect of fishery development, with its fishery resource production and output values.

Zone-4

This zone is characterized by low fishery resource development. There are two taluks in this zone, namely, Tiswadi and Canacona taluks. Both the taluks are situated along the coast. There is great scope for fishery resource development for the people in this zone. Comparatively, these two taluks are lagging behind in respect of fishery resource production and output values. Tiswadi taluk enjoys 4th position in production, output values and traditional boats, 3rd position in mechanized boats, and 6th position in percentage of fish catching centres and the number of population engaged in marine fishery activities.

However, the Canacona taluk records the lowest ranks, that is, 6th in output value and production, 3rd in traditional boats and fisherman population, 2nd in fish catching centres and 5th in mechanized boats. Thus, there is enormous scope for development of fishery resources and activities as well. During 2009-10, Pernem taluk entered low development zone.

Zone-5

The last zone represents the very low fishery resource development for a period of 2001-02 to 2007-08. Pernem taluk lies in this zone. Pernem taluk enjoys 6th position in traditional boats, 5th position in fish catching centres, production and output values, 4th position in fishermen population and mechanized boats. Fishery resource development is very limited due to various problems, i.e., lack of proper infrastructure in this zone. Hence there is great scope for fishery resource development. People of Pernem taluk need to be encourage to take keen interest in fishery activities and to avail themselves of benefits of government and private agencies. During 2009-10, Tiswadi taluk has entered very low fishery resource development zone with a meager production and output values. Tiswadi taluk needs to be given proper attention by the Directorate of Fisheries and Fishermen community, to study the cause of decline in the productivity of fisheries, despite the presence of a good number of mechanized and non-mechanized boats as well as active participation of large fishermen population.

Quepem taluk is not a fishery taluk in its true sense, as it has rocky coastal belt, approximately one kilometer. In fact, it has elevated coastal topography wherein land and water interchange is very limited. Lack of infrastructure, i.e., fishing ramp, jetty, cold storage, auction shed, etc., has affected fishery activities adversely and naturally led to under development. Only a few fishing activities are carried out at the subsistence level only for domestic consumption by the people of this area.

Hence, the hypothesis the marine fisheries production is higher than the inland fisheries has been tested with literature and confirmed the hypothesis.

CHAPTER–4
MARKETING OF FISHERY RESOURCES

4.1 INTRODUCTION

In earlier days the term marketing of fish meant buying and selling of fish at the lending centers. After the Second World War marketing of fish has taken a new role in business activity. The fisheries have not become highly industrialized in all fishing nations. The new fishing techniques have been adopted to sell more fish. The modern fish marketing system lays emphasis on meeting the existing demand of fish, besides tapping the potential demand in the important markets. The marketing of any produce mainly depends upon the availability, consumption and demand. In Goa, traditional system of fish marketing is adopted. Modern marketing system as well as the fish marketing is normally done at the collection centres, which are mainly situated in the area of fish landing.

Fresh fish may be sold daily in fishing ports located near fishing areas. However, fish and fish products to be sold in distant markets must be processed first to prevent spoilage. Many fishing crews sell their catches to processors at auctions after fishing trips. The price which a catch commands depends on the supply of fish at the market and demand for it. A fishing crew does not know in advance what a harvest it will reap, if it is sold at all. Sometimes the processor places orders with the fisherman before fishing trip. At the same time, both sides agree on the price to be paid for the catch. Sometimes, fish may not be consistently demanded by customers due to its peculiar smell. It is the job of fishery technologists to prepare the odorless fish products for wider consumption. Thus, the fish marketing objective should not only be catching and selling of fish but, also seeking a wider scope for exploitation, production, distribution, preservation, packaging and transportation, etc. of fish, in addition to effecting direct sales by avoiding middleman.

According to the New Oxford Dictionary (2007) Encyclopedia, "a fish market is an open space in town, village, city, etc. where people gather for the purchase and sale of fish".

4.2 FISH MARKETING

The fisherman visits the fishing grounds, and if he is fortunate to strike a catch, tends to bring the produce to the nearby market for sale as soon as possible. In some cases the fish catch may be of good size and variety acceptable to the customers while in some cases it may be poor. There are no effective measures for regulating the catch or forecasting the varieties of fish that would be landed. The quantity of fish is also uncertain. This therefore, creates shortage, which affects the fish marketing and pricing of fish.

The different type of fish markets are as follows:

1. Domestic Marketing

The decline in the share of production of domestic fresh fish can be attributed mainly due to factors such as the nature of mechanization and lack pf proper marketing infrastructure. Growth in the mechanization field has been dominated by trawlers which mainly concentrate on exportable varieties, and low value fish brought along with shrimps has a low consumer preference, adding very little to fish suppliers for domestic fresh fish marketing. In the case of purseiners, the landing was usually very large in quantity, and the existing marketing resulting in large scale diversion of drying yards. While the share of domestic fresh fish in total landing decreased by half, the share of dry edible fish and fish meal virtually doubled. The varieties used for dry edible fish are mostly those which do not have a ready market in fresh form in nearby places or, which are landed in large quantities and the fresh fish marketing infrastructure could not handle them.

2. Export Marketing

Most of the exporting units get the sea food directly from the selected trawlers, retained on contract along various centres of Goa. They have their suppliers who have a direct linkup with the fisherman. Thus, the suppliers are the intermediaries on commission basis, who arranged the network and form a proactive link-up. In the beginning of the fishing season, i.e., before retreat of the monsoon season, these suppliers are approached by all the fishing units. The advances are given to the boats committed by the suppliers. When the trawler brings the fish they unload it in boxes of the company. They are then weighted and transported to the processing centres. Most of the fish which is exported outside India is either in fresh or processed state. The fish when brought to the processing plant are separated according to their size, washed clean and packed in polythene bags, and are kept in chilled rooms. The importers, who want to purchase the frozen fish, send their own persons to inspect the fish and to place orders. Most of the high quality fish, like the Ribbon fish, Mackerel, Tuna Bill fish, Pomfret, are exported to foreign countries. Goa exports mostly frozen fish to China, Japan, USA, Middle East, Spain, France, European Union, South East Asia and other countries.

The foreign exchange earnings from export of fish and fish products have been steadily rising during the years. The export earnings of this commodity can be more than double within 3-4 years.

3. Primary Market

This is a marketing place at the catching point, usually in a rural area. Fish collectors/assemblers commonly known as Mahajans or Aratdar/Mahajans produce fish from the catchers, with the help of local brokers called dalals who get a profit margin or commission from the Mahajans. Part of the catch is also locally sold by the catchers/farmer or by local retailers.

4. Secondary Market

The collectors bring the fish from the primary market to the landing ghats, usually to the nearest Thana market or at a place well linked by rivers, road or rail transport. The Mahajans sell the fish here to the distributors known as beparies, generally with the help of the aratdars, the commission agents.

5. Higher Secondary Market

The beparies transport the fish to the nearest city/town by markets by road, rail or boat. There are mainly distribution markets, and here the beparies sell the fish to another set of distributors known as paikars, again with the help of aratdars.

6. Final Consuming Market

On purchasing the fish at the higher secondary market, the paikars sell the fish to the retailers. There are two channels of retailing markets set up permanent stalls, or set out with the fish on their heads or in tricycle (rickshaw) vans, to sell them at homes. Other retailers take the fish to suburban places or to the villages around the city/town.

In the course of marketing at all these levels the collectors or distributors carry out the function of handling, clearing, sorting, icing, preservation and transportation at his own cost as far as possible. Expenses on such accounts are deducted from the bills or sellers.

The success of the marketing of fishes depends much upon how efficient the fish marketing system is. Simply put, fish marketing finds a good market to sell fish and its products, serving as a direct link between the trader and the consumer. Its utility and efficiency lies in safeguarding the interest of not only the traders but the consumer as well. It must satisfy both of them. Also the trader must not be allowed to feel satisfied with marginal returns, average production and medium quality. Fish marketing here has an important role to play. It must induce the trader to endeavor for maximum production, best quality of commodity and its timely supply at reasonable cost. This can be achieved only when fish marketing is able to offer a really encouraging market in and outside the country. Its role in earning foreign exchange and thus promotion of the economy of the country need not be emphasized as also its effectiveness in solving to some extent the food problems that countries faced with storage.

3H
3H

Functions of Fish Marketing

The following are the significant functions of fish marketing:

1. To know when, where and what is the demand position of the consumer market,
2. To assess the demand of any one place as to its amount, kind and time with regard to supply of with hold fish,
3. To regulate supply in such a way as to enable stabilization of price in periods of bumper landing or extremely poor landing of fish. The distribution system at all levels has to be supervised efficiently to prevent occurrence of local artificial scarcity, and consequent price hike, as also any flooding of commodity, and consequent dangerous price fall,
4. To determine the proportions of fish landed for diversion to various agencies, dealing with preservation, storage, caring yards, canning industries and manufacturing units of various products in accordance with requirements of local and foreign markets,
5. To establish standards of quality control at various levels of storage, prevention and production of local and foreign markets,
6. To prevent loss due to wastage,
7. To keep a strict vigil on unscrupulous elements and their activities in the trade. Middlemen, co-operative societies, moneylenders, hoarders and retailer engaged in the trade to be controlled for their harmful activities or functioning,
8. To educate the people about the importance of food value of fish and to popularize, through publicity, in them the habit of eating fish,
9. To explore possibilities of export and import, thereby to help the country to earn more foreign exchange and cooperate in international trade.

Marketing of Cultivated Fish

It is useful to know the growth, age, sex and behavior of certain individual fish of any particular species cultivated in a single pond. The identification of these individual fish is made by marketing. The first will permit recognition of a determined group of individuals, and the second the identification of an individual fish.

a) Collecting Marketing

A simple way of collecting marketing is by clipping a fin or part of a fin with scissors. This technique can be used for nearly all species and sizes of fish. Generally speaking, experienced personnel can recognize the marketing without too much difficulty even after one growing season. The re-growth of a clipped fin is much more pronounced, as it is cut near its point of implantation. It is important not to injure the fish while allowing it to bleed when the fin is clipped. A partially clipped fin after complete re-growth can still be recognized, thanks to the deformed fin rays at the point of clipping. This deformation can be recognized more easily by transparency. Numerous tests made in ponds have shown that fish scarcely suffer from this marketing technique.

Fish Catch From Trawl

Either the right or left ventral or pectoral fin or the top or lower part of the caudal fin can be clipped. This permits several groups of individuals in the same pond to be distinguished. It is also possible to increase the number of groups by clipping two fins.

There are other systems of collective marketing such as tattooing or using colored subcutaneous injections, but these are less practiced than those given above.

b) Individual Marketing

Individual marketing is generally by metal or plastic numbered tags. These can be obtained in several models. They are fixed to different parts of the body. Some tags are fixed to the jaw or to the gill cover though these are only suitable for larger fish and species with a pronounced lower jaw such as trout or perch or with a sufficient strong gill cover such as in carp. These tags often cause injury, and those attained to the jaw can hinder feeding.

For sufficiently large fish a tag fixed by a metal wire or some synthetic material passed through the muscular fibers near the dorsal fin is often used. Losses of tag can be rather high. It is often nylon thread fixed to the large spiny ray of the dorsal fin of carp.

Another method of marketing is to brand the skin of the fish. This applies only when fish have scales or no scales such as mirror carp or leather carp as well as fish with very small scales. This can be done with a silver nitrate pencil or a heated metal wire such as a copper wire heated on a small portable stove, or a platinum or chrome nickel wire heated electricity. This method permits collective marketing, if only one sign is used, a cross for example. It also permits individual marketing, if number is used. This method is simple and rapid and practically without danger for the fish. Collective marketing and above all, individual marketing is greatly done by the use of an anesthetic. There are several anesthetics available, and MS222 or propoxates are commonly used at present. First, a stock solution with 20gm of MS222 per liter of water is prepared. The fish are then placed in a bath containing 5cm of the solution, per liter of water. Propoxate may be used for the anaesthetesia of salmonids at a level of 1 to 2ppm, and for certain other freshwater and marine species at a level of 1 to 4 ppm of active ingredient. The moment the fish turn on their sides they can be removed from the bath and marked.

Factors Influencing Fish Marketing in Goa

Generally, the fish marketing is largely governed by various factors in the – study area, i.e.:

1. Great Demand for Fish and Fish Products

Fish is the staple food of Goans, more than 90% of total population consumes fish on daily basis. The density of population is high in Goa. Therefore there is a great demand for fish and fish products throughout the year, and a large portion of the total catch is marketed within Goa itself. During the lean season fish is imported from nearby states to meet the rising demand, i.e., Kerala, Maharashtra.

2. Climate

The state has been experiencing balanced climatic conditions throughout the year. Hence the climate of Goa is divided into three seasons, the rainy season, summer season and the winter season. During the rainy season, the conditions are usually unfavorable, when the sea is rough, and not suitable for fish catch. From September to March, conditions are very ideal for fishing activities, benefiting large number of fishermen community in the study area.

3. Transport

Transport development is rapidly taking place in Goa. The state is well connected by National highway, State highway, a number of village, talukas and district roads. Fish being a perishable commodity, needs to be quickly marketed and sold. Flexible network of transport enables quick and safe disposal of fish products in Goa. The density of vehicle population has been increasing year after year to meet the needs of ever-growing population.

4. Availability of Catch

A large number of people are involved in fishing activities in Goa, due to which huge amount of fish is caught commercially and locally by the fishing community. Effective means of transport allow the fish to be distributed to different centers quickly. The excess quantities of fish is sold to the neighboring states, and even exported to other countries.

5. Marketing Centers

Goa has well established marketing centres both at the urban and the rural levels. Goa has well established fish markets and also large marketing centres for fish and fish products. Due to efficient means of transport and convenience large quantity of fish and fish products is traded all over Goa.

6. Infrastructure Facilities

Infrastructure facilities are well established in Goa. The study area is well endowed with basic infrastructure facilities to encourage fish tracking activities on large scale, i.e., transport, cold storage, ice factories, auction sheds, net mending sheds, etc.

7. Capital

Goa has many financial institutions. A large number of Nationalized Banks, Co-operate Banks, Credit Societies, etc., are operating here. Almost every village has some or the other means from where capital can be derived. Fishermen and traders borrow capital on different terms and conditions to manage their activities. Government also provides financial assistance by way of subsidies, grants etc. through Nationalized banks.

8. Influx of People

The state is well known for tourism. Hence, there is large movement of people from all over the country and the world as well. This has created a situation of rapid growth of hotels, restaurants and shacks in different parts wherein fish is consumed largely by the tourists giving rise to fish marketing on commercial scale.

4.3 MAJOR COMPONENTS OF FISH TRADE

The following are the major constituents which influence fish trade greatly in the study area.

1) Role of Fishermen and Marketing Intermediaries

Earlier the fishermen utilized small boats and they fished close to the land on the reefs or inshore banks. In the past, industry had a simple marketing structure where fishermen themselves marketed the fish at landing sites or from the back of open trucks, or a small number of vendors sold the fish to the local people.

Now the fresh fish passes through one to eight hands before it reaches the final consumer located at a multitude of places in rural, urban and metropolitan centres. However, the bulk of fresh fish passed only through one to three hands. The fresh fish was sold by fishermen directly to retailers or vendors at the producing centres. The direct sale by fishermen to retailers/vendors often constituted the totality of family enterprise where women from the same families retailed the catch of their own boats. The market for fish at most of the centres was oligopolistic.

The fishermen either directly sell fish to wholesalers and commission agents through them to retailers. The commission agents, wholesalers - cum agents and wholesalers do not sell directly to consumers except to bulk buyers such as hostels, restaurants, hotels and exporting companies. Otherwise, they sell to retailers and vendors who come to the wholesale market and purchase fish either at auction or directly. When the production of fish is very low, it is directly sold to consumers by fishermen.

In many places, the buyers may be the wholesalers or fish merchants or middlemen. The fishermen send fish to the commission agents as consignments. The commission agent auctions the fish, and the gross sale proceeds are remitted to the fishermen after deducting various marketing charges, like clearance and cartage charges, packing and ice charges, octroi and termination taxes, etc.

When the wholesalers or agents buy fish directly from the fishermen, they enter into a contract with the exporting units, hotels, etc. During the rainy season when there is less fish, the wholesalers sell the iced fish which he has purchased or kept in the freezers during peak season. He sells it to the fisher folk, hotels, etc., at a very high price, thereby making a lot of profit for himself. Thus, the role of the fishermen is totally passive, as they play no part in the marketing functions, thus allowing the numerous middleman to obtain huge benefits at their cost.

Similar thing happens to the processors too. They sell most of their fish products to fish brokers in large cities at a very less price. The brokers, in turn, sell the products to restaurants & food stores at a high price which includes their commission. The food stores and restaurant charge it still at a higher price and by the time it reaches to the ultimate consumer, the price is four times more than the actual cost of the product.

Thus the process of marketing of fish and fish products does not favour the fishermen. It favours the middlemen only.

2) Price and Price Determination of Fish

The fluctuation in fish price is very prominent. The changes are too frequent to predict any trend. There may be one price in the morning while another in the evening in the same market. Sometimes prices change at short intervals of time that is even, by the minute. The price rise is not only due to the sudden supply and demand of particular variety of fish, but, also due to prices of other varieties in the market.

Definitely, the perishability tendency of fish, has a significant role to play in determining the price of fish at the markets. The fish price often vary from spot to spot in one fish market where many auctions are taking at a single time. When fresh fish reaches early in the morning, a high price is quoted, and it generally falls as the day advances. The uncertainty of supply and demand plays a great role in price determination. In the fish market, at the time of auction, the pace of selling fish is so fast that it cannot permit news of low prices. There is a great price risk because of high perishable tendency of fresh fish. The fish markets are having certain timings. Generally, the auctioneer starts the bidding price based on the experience. It is also observed that some buyers in some instances purchased one type of fish at a higher price from one particular agent, while the same was bought at a lower price from a neighbouring agent, indicating a variation in prices. Time factor plays a prominent role in almost all fish markets, because there exists some anxiety with the buyers to buy the fish as early as possible for the fear that there would be none left later. Huge supply of fish normally means that there is unchanged demand and buyers will ask for fish at a very low price.

Many a time, specially in the state of Goa, when Hindus celebrate their feast like Shravan, Ganesh, Dussehra, etc., they stay away from fish and this is the time where fish is sold at a lower price.

The Goan housewives can doubt that the price of fish has shot through the ceiling in recent years. The poorest among society will be hardest hit. A price rise of fish will take a bigger chunk out of their already small wage. And also generally they are the most lacking in proteins. The reasons why the price of fish increases are as follows.

1. The fresh fish caught by the trawlers is sold to other states of India to convert into fish meal and manure. A decade ago this same fish caught by the traditional fishermen was utilized by the locals.
2. High quality fish, like pomfrets, king fish, squids, tiger prawns, etc., are exported abroad substantially.
3. The increasing tourism industry has pushed the price up by paying top dollar for the prize catch. Most of the fish caught is sold to 5 star hotels for a higher price. Only the seasonal fish, like mackerels, sardines, pedve, kampi, velli, etc., reach our markets for the Goan inhabitants in large quantities, affecting Goan consumer by and large who is forced to pay a high price for poor quality fish for his daily consumption.

3) Packaging

Packaging is an essential part of processing and distributing foods. In the olden days, fish products were preserved by sun-drying and pickling. Now preservation of fish products is done by controlling bacterial and chemical action mainly by freezing or by freeze drying and keeping them in good condition by the use of modern methods of packaging. A faulty packaging will undo all that a food processor has attempted to accomplish. Packaging should meet the size, shape and weight requirements of the buyers. Their appearance should be in such a manner that they can easily attract the buyers. We shall learn how packing of live fish and fish products is done in details as we proceed in this matter.

The fish has to be packed neatly with ice in suitable containers for transportation. Fresh fish is wet and liquid will be oozing out from the fish. Even very fresh fish has bad smell and slime will give stinking smell. Ice in the iced fish will melt and the ice melt water has to be let out. Otherwise the fish will get adversely affected.

In India, baskets have been the traditional containers used for fresh fish transportation. They are still used largely by the trade. Baskets made of split bamboo, harhar, palmyra leaf mattings, etc. are employed to pack fish for transportation. Bamboo baskets are cheap and clean when new, disposable single service type containers. They are of various sizes 40 kgs. – 100 kgs. of fish. Their price starts from Rs. 30/- to Rs. 80/-. Sometimes these baskets are given a gunny lining all around after packing the fish, which gives the basket additional protection and also absorbs some of the ice melt water on the way. Such baskets are generally used for transporting iced fish by bicycle, handcarts or even as head loads to nearby places. But the demerits of this packaging is that these baskets get deformed soon, it causes bruises on the fish skin due to rubbing against the rough surfaces of the basket and these cuts on the fish bring about easy spoilage. But they are still used in trade.

Deal wood/plywood boxes are generally used in western part of the country for transportation of fish, especially for long distances by wholesalers and commission agents.

Dry fish is usually irregular in shape leading to difficulty in assembling in a neat package. Packaging of dried fish is a neglected area in marketing. Commonly used packaging in the country side is either old newspapers or dried leaves. However, packaging materials tried for commercial use in India are:-

1. Waxed corregulated cartons;
2. Deal wood or ply wood boxes;
3. Bamboo baskets;
4. Gunny bags;
5. Dried Palmyra or coconut palm leaf mats;
6. Multi wall paper sacks.

Package of raw frozen fish portions or fillets is mostly of polyethylene, either as premade bags or wraps in form-fill seal machines, moist or hot-melted coated paper board cartoon with or without a waxed-paper overwrap. Polyethylene accounts for the biggest proportion of the plastic used in most commonly, since it possesses many desirable qualities such as transparency, water vapour, impermeability, heat sealability, chemical, and it is fairly economical.

At the freezing centres, the finished products are conveyed to packing area. At this point, the product is weighed and according to the grades/size and identification on the production date, code is placed in the polythene slab of the frozen sea food and placed into the master carton as required by the customers. These cartons are stored in a clean, dry and hygienic storage atmosphere as the packaging media should have good property to moisture loss, dehydration and oxygen does not impart odour, and prevents flavour loss and withstands low temperature storage.

For pickling of fish, the package mostly used is tin cans. They are mostly used to protect metal can from corrosion by the food. The can must be filled with a correct weight of material and then sealed. The main objective of this process is to obtain air¬tight seal between the cover and the body of the container so that spoilage agents cannot enter the sealed container after the canned fish has been sterilised. The can then has to be labeled. This label must bear the name of the product, net contents and other specific information as required. The object of storing is to hold the canned fish under conditions that do not alter the quality of the fish or the appearance of the container. The processed can must not be cased hot, because the loss of heat by radiation from the cases is slow and this will cause injury to quality. The packaging cost of these containers works out to almost 18 to 20% of the retail price of the product starting from C.S.T. 4%, tax 1-5% sales tax 8% and surcharge 12%. Similar structure exists in every state but in varying degrees.

4.4 WHOLESALING AND RETAILING

Market refers to any specific place in a town or in a city where buyers and sellers of specific goods and services meet to buy and sell the goods and services.

The early records of the human history show very clearly that salesmanship existed earlier in primitive form. Since barter system suffered from certain difficulties such as lack of double coincidence of wants, etc., buyers and sellers were in need of an article which would be double and acceptable to the community as a standard of value. In course of time the difficulty of buyers and sellers were solved when coins and paper currency came to be accepted as a medium of exchange.

Barter: selling in ancient times mainly consisted of barter. It means exchange of one commodity for another.

Local markets: In early times, villages and towns were self sufficient economic units. The needs of the people of villages and towns were met by petty shop keepers and by the sellers who went around from house to house, who are known as Hawkers and Peddlers. They are familiar figures even today. In this way salesmanship arose in crude form under barter economy. However, salesmanship and commerce were occasional occupation. It was only at the village economy stage that both salesmanship and trade got firmly established. The buying and selling activities were conducted in a regular continual manner at the town economy stage. The introduction of money and middleman led to an ever increasing role of marketing and distribution.

Wholesalers: The wholesaler is the first link in the chain of distribution of goods from the producer to the ultimate consumer. Since the producer and consumer are located at distant points, some intermediaries are required to link them with one another. This link is provided by the wholesaler and retailer. Of these two, wholesaler is in direct contact with the producer, while retailer enjoys the single position of being in direct contact with the number of consumers. Wholesalers and retailers are essential for the distribution of goods. (The wholesaler can be defined as a merchant who acts as an intermediary between the primary producer, manufacturers or importers on one hand, and retailers or consumers, on the other) Bari Mulay (1988).

In fishing occupation, wholesalers are normally the owners of the fishing vessels who sell the fish on wholesale basis.

Types of Wholesalers

1. Large scale wholesalers:- In fishing occupation, wholesalers are, normally the owners of the fishing vessel, who sell the fish on whole-sale basis. However, the entire catch of the fish from the owner of the fishing vessel is not purchased by a single person but by several persons known as distributors.
2. Medium wholesalers (distributors):- They are the ones who purchase fish from the trawler owners, and sell it to other wholesalers. They purchase fish in bulk from the trawler owners, and sell it to other small wholesalers.
3. Small wholesalers:- They are those who purchase the fish from the medium wholesaler and sell it to the retailers. They go to the markets with small pickups, and sell fish to the retailers in small quantities as demanded.

Retailers

Retailing is a kind of the internal trade, which is concerned with the sale of goods to the consumer in small quantity (Bari Mulay, 1988). A merchant engaged in this type of selling is called retailer. He is the last link in the chain of intermediaries connecting the producer, the manufacturer, on one hand, and the ultimate consumer, on the other.

The most important characteristics of retailer is that he comes into direct contact with consumer, and renders a number of services to him. Here retailers may be defined as the fish vendors who purchase the fish from wholesalers in large quantities at economic prices, and sell it to the ultimate consumers.

Types of Retailers

1. Large scale retailers:- They are retailers who purchase fish from the wholesaler, and in turn, sell it to the ultimate consumer. They purchase in large quantities, and their customers are mostly the hotel owners who buy in large quantities. They sell the fish in the market itself and do not go from door to door.
2. Ordinary retailers:- They are hawkers and peddlers who sell fish at the door step of consumers by going house, to house marketing their fish. They either purchase from large scale retailers or sell their own catch. Many people prefer to buy from such ordinary retailers as they come with fish at the door step and save their time to go to the markets to purchase fish.

4.5 INTERNATIONAL TRADE IN MARINE FISHERIES AND FISH PRODUCTS

Marketing of fish is an important aspect of international and domestic trade. More than 90 percent of people consume fish as their main protein source and regular diet in the study area.

Marketing of fish and fish products at the international level started in the year 1975. The frozen prawn export packed up, but under stiff competition and changing international tariffs, there was a setback. After 1986, export of marine products is a significant contributing factor in the state economy. Before 1988, export was quite negligible, the real export of fish and fish products commenced from the year 1990-91. The state is earning valuable income by exporting a variety of fish products and raw fish to various countries of the world, i.e., Malaysia, China, South Korea, Singapore, Mauritius, Japan, UK and Spain.

The following table depicts the trend in export of marine products and fish from Goa.

Table-4.1 : Trend in Export of Marine Products and Fish from Goa During 1988- 2008.

Year in M.Tons	Fishery Products Growth Rate	Percentage of in Rs. Crores	Total Value
1988	1234	-	4.41
1989	893	-27.63	1.74
1990	1431	37.59	2.44
1991	4288	199.65	8.21
1992	4429	3.28	17.61
1993	9566	115.9	45.54
1994	17193	79.73	65.66
1995	13474	-21.63	67.53
1996	11908	-11.62	60.03
1997	14284	19.95	66.08
1998	6175	-56.76	29.10
1999	9054	46.62	34.91
2000	10732	18.53	33.59
2001	7714	-33.33	30.07
2002	15594	102.15	56.18
2003	10288	-34.02	32.79
2004	8856	-13.91	39.09
2005	11001	24.22	61.78
2006	14117	28.32	72.99
2007	17531	24.18	88.49
2008	21328	21.65	196.89

Source: Directorate of Fisheries, Government of Goa (2009-10)

From the above table it appears that the production and output values fluctuated over the years from 1234 m.tons in the year 1988 rise to 14288 m.tons in 1997, which shows gradual growth during the decade in respect of export of marine products and fish. After that, there was sudden decline, reached to the lowest 6175 m.tons by the next year 1998, witnessed 56.76 decline due to less international demand for the fish and fish products. Subsequently, the output values have come down to 29.10 crores. From 1999 onwards there was sharp increase in exports, i.e., 9054 m.tons to 15594 m.tons in 2002 and further it steadily went upto 21328 m.tons valued at Rs. 196.89 crores by 2008-09. This has been a remarkable growth in the export of marine products and fish during the last two decades in Goa. The demand for fish products has been recently increasing by a number of countries in the world.

The following table shows the species wise export of marine fish and fishery products 1999-2006.

Table-4.2 : Export of Marine Fish Products (Quantity in M.Tons)

Sl.	Items		1998	1999	2000	2001	2002	2003	2004	2005	2006
1	Fr. Shrimps	Q	188	85	33	51	-	-	18	-	-
		V	406	276	103	105	-	-	56	-	-
2	Fr. Cuttle Fish	Q	1197	1188	561	216	691	882	766	641	628
		V	898	613	245	74	364	298	516	383	598
3	Fr. Squids	Q	668	488	140	393	49	0	22	-	22
		V	511	407	191	288	22	0	21	-	18
4	Fr. Fresh Fish	Q	0	0	0	0	-	0	27	-	-
		V	0	0	0	0	-	0	14	-	-
5	Fr. Tuna	Q	0	0	0	394	-	0	594	-	-
		V	0	0	0	191	-	0	196	-	-
6	Ribbon Fish	Q	2288	1980	5622	1272	8238	4692	3461	4243	5961
		V	490	470	1403	356	2638	1292	1045	19683	1618
7	Indian Mackerels	Q	819	2037	3401	2989	2383	1990	1227	1443	2358
		V	168	500	981	1053	1021	839	453	5778	1251
8	Reef Cod	Q	6	876	220	1961	2341	2098	370	430	287
		V	1	247	62	606	782	553	99	127	144
9	Fr. Seer Fish	Q	208	44	128	200	31	166	286	727	132
		V	166	22	969	135	78	91	193	604	304
10	Fr. Assorted Fish	Q	801	2376	987	235	177	344	474	380	-
		V	270	958	359	199	657	157	282	177	-
11	Other Fishes	Q	0	0	0	0	134	116	1610	3138	4729
		V	0	0	0	0	56	43	1032	2347	3366
	Total	Q	6175	9074	11092	7711	15594	10288	8855	11001	14117
		V	2910	3493	4313	3007	5618	3273	3907	6178	7299

Source: Directorate of Fisheries, Government of Goa (2010).

From the above table it appears that various species of the fish and fishery products have been exported to the needy countries of the world.

The notable feature is the export of fish species like fresh shrimps, fresh cuttle fish, fresh squid, fresh tuna, ribbon fish, Indian mackerels, reef cod, fresh seer fish, fresh assorted fish and other fishes. The trend in export of marine fish and fish products is fluctuating over the years. It has been observed that there is no constant demand from foreign countries for increased supply. It is clearly evident that the year 2002 saw steep rise in export of marine fish and fish products amounting to 15549 m.tons valued at Rs. 56.18 crores compared to 2003 where the total exports were 10.288 m.tons valued at Rs. 32.79 crores. Similarly, in the year 2008-09 the exports amounted to 21328 m.tons valued at Rs. 196.89 crores.

The fisheries department pointed out that except for prawns most of the exported fish like ribbon fish, baim fish and squids are rarely bought by the Goans. Hence the export of marine fish and fish products does not affect the local market.

The following table displays the unit-wise manufacturer, exporters, products which are exported to different foreign countries of the world.

Table-4.3 : Unit-wise Manufacturer Exporters Products which are Exported

Sl. No.	Company /Unit	Type of Commodity	Fish Processed Qty. Metric Tons per day	Export of Fish Product Estimated Capacity Aug. 15 – Nov. 15 = 77 days Qty. Metric Tons	Percentage of Total
1	Rahul Foods, Danjiela, Old Goa, Tiswadi	Frozen marine products	89	6853	18.59
2	Corlim Marine Exports, Sancale Industrial Estate, Zuari Nagar, Mormugao	Frozen marine products	34	2618	7.1
3	M/s. Ulka Seafoods Pvt. Ltd., II Tiverm Marcela, Tiswadi	Frozen marine products	60	4620	12.53
4	Goan Bounty, Dulapi Corlim, Ponda	Frozen marine products	73	5621	15.25
5	M/s. Ulka Seafoods Pvt. Ltd., II Tiverm Marcela, Tiswadi	Surimi product	35	2695	7.31
6	M/s. Seahath Canning Company, Margao IE Salcete	Canned products	14	1078	2.92
7	M/s. Atlas Fisheries Pvt. Ltd., Bainginim, Old Goa, Tiswadi	Frozen marine products	55	4235	11.49
8	Quality Exports, Cuncolim Goa, Salcete	Frozen marine products	57	4389	11.91
9	M/s. Quality Foods Cuncolim Goa, Salcete	Frozen marine products	61.5	4735	12.85
	Total		**478.5**	**36844**	**100**

Source: Field Survey by Researcher (2010)

From the above table it appears that there are nine companies/units involved in fish processing in order to produce variety of marine products. Among various units, the Rahul Foods situated at Danjiela Old Goa, Tiswadi taluk ranks first in respect of fish processed capacity of 89 metric tons per day accounts for 18.59 percent of total. The unit specialized in processing of frozen marine products to cater to the needs of foreign countries. Rahul Foods can produce maximum upto 6853 metric tons over a period of three months during peak season of fisheries from August 15th to November 15th.

Goan bounty is an important producer of frozen marine products ranks second in exports of the same. The unit produces about 73 metric tons per day, which accounts for 15.25 percent of total, and also it undertakes processing of marine products during peak season of fisheries in the study area. The estimated production of the unit is around 5621 metric tons, followed by M/s Quality Foods located in industrial estate of Cuncolim, Salcete taluk ranks third in total manufacturers, exports. The unit produces about 61.5 metric tons per day and its estimated capacity of production, i.e. 4735 metric tons during peak season of fisheries. M/s Quality Foods, which accounts for 12.85 percent of total export output and values.

The remaining manufacturer exporters are M/s Ulka Seafoods Pvt. Ltd., produces 60 metric tons per day of processed fish products (12.53%). Cuncolim produces quality exports about 57 metric tons of marine products per day (11.91%) of total.

The Corlim Marine Exports, unit which is located at Sancole industrial estate, Zuari Nagar, represents comparatively less processed fisheries output, that is, 34 metric tons per day can produce maximum upto 2618 metric tons of marine fish products accounts for 7.10 percent of total in the study area. Thus it has been viewed that all the manufacturer exporters units making substantial contribution towards development of international trade in respect of marine fish and products is largely benefiting Goa's economy.

4.6 NEAREST NEIGHBOUR ANALYSIS

In the regional studies the study of spatial as well as temporal variations in distributional pattern of fish landing centres, settlements and market centres is of great importance. According to Watson (1955), geography itself is a discipline in distance, and according to Cole and King (1968), geography is the science that is mainly concerned with the distribution of element that occur on the earth surface and with the variations of distributions through time and space.

Table-4.4 : Distribution of Fish Market Centres in Goa

Census Towns Population in North Goa	Nos.	Census Towns Population in South Goa	Nos.
Pernem	5289	Mormugao*	97154
Parcem	4324	Chiclaim	7604
Siolum[0]	10318	Sancole[0]	15604
Colvale	5475	Margao*	78382
Mapusa*	40487	Benaulim[0]	10158
Saligao	5559	Nanelim[0]	11014
Calangate[0]	15783	Davorlim[0]	10929
Candolim	8604	Saojose Deareal	8351
Reismagos	8708	San Vordem	4833
Penha defranca[0]	15377	Curchorem[0]	21407
Soccoro	10174	Sanguem	6173
Aldana	6588	Quepem[0]	12573
Bicholim[0]	14913	Chinchinian	7033
Carapur	5339	Cucolim[0]	15866
Sanquelim[0]	11194	Canacona[0]	11901
Valpoi	7917		
Pale	5641		
Chimbel[0]	11984		
Panaji*	70078		
Calapur[0]	11830		
Bambolial	5785		
Goa Velha	5395		
Bandora[0]	12267		
Quela	5456		
Ponda[0]	17713		
Curti[0]	13179		

Index:

Major

* above 30,000

[0]Minor

10-30,000

V. Minor

0-10,000

Source: Directorate of Census Operations, (2001)

Spatial Distribution of Fish Market Centres

$$Rn \quad \frac{\overline{Do}}{\overline{De}}$$

$$\overline{Do} \quad \frac{\text{Total Measured Distance}}{\text{Total No. of Settlements}}$$

$$\frac{202.2}{43}$$

$$4.70$$

$$\overline{Do} \quad 4.70$$

$$\overline{De} \quad 0.5\sqrt{A/N}$$

Where as

A Area

N No. of Settlements

$$0.5\sqrt{\frac{3702}{43}}$$

$$0.5 \quad 9.27$$

$$4.63$$

$$\overline{De} \quad 4.63$$

$$Rn \quad \frac{\overline{Do}}{\overline{De}}$$

$$\frac{4.70}{4.63}$$

$$1.01$$

$$Rn \quad 1.01$$

GOA – SPATIAL DISTRIBUTION OF FISH MARKET CENTRES

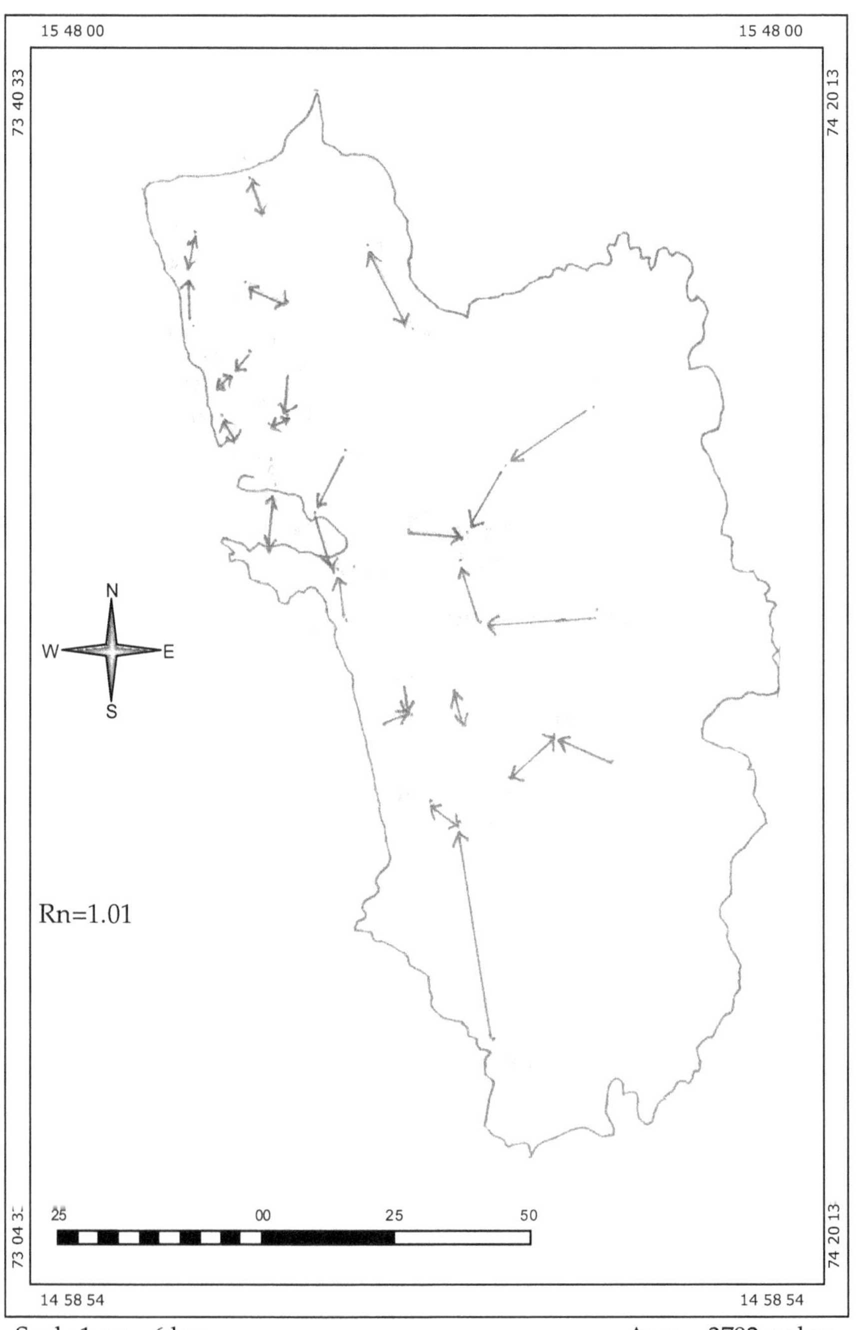

Scale 1cm = 6 km

Area = 3702 sq.km.

Map-4.1

The formula for nearest neighbor analysis is

$$Rn \quad \frac{Do}{De}$$

Whereas,

$$De \quad \frac{1}{2}\sqrt{A / N}$$

Whereas,

Do = The mean nearest neighbour distance as observed.

D : = Distance from its points to its points to its nearest neighbour

The figure shows the results based on the nearest neighbour analysis of fish market centres in the study area. As per the study spatial distribution is concerned, the calculated value is 1.01, which is described as perfectly random. Thus the fish market centres are randomly distributed in the study area.

From the above table, we learn that there are four major fish market centres located in different parts of Goa, i.e., Mapusa from Bardez, Panaji from Tiswadi, Margao from Salcete and Vasco from Mormugao talukas. These towns have more than 40,000 population, wherein fish customer flow is very high that is, above 10,000 per day in each fish market centre to do necessary purchases.

There are eighteen minor fish centres located in various parts of north Goa and southern parts of Goa. The settlements have population in the range of ten thousand to thirty thousand and the fish customer flow in each fish market centre is observed to be between five thousand to ten thousand.

Fish market centers of lowest costumers flow in the range of zero to five thousand or below five thousand can be termed as minor, as road side fish markets are developed in these towns. There are twenty one towns with a population of less than ten thousand. Thus it has been observed that the fish consumption is very high in major towns / cities of the study area due to high density of human population, which is largely influenced by the growing tourism industry, service sector, industrialization, construction industry and other important economic sectors in the state.

Moderate level of consumption of fish could be seen in minor census towns of the study, area and very low consumption has been observed in sparsely populated regions of Sanguem, Quepem and Sattari talukas of Goa. Naturally, the supply of the fish is limited in these areas as they are mountainous regions covered by dense forests and rough topography.

Table-4.5 : Spatial Analysis of Marketing of Fishery Resources of Goa

Code / Sl.No.	Taluka	Area / Sq.km.	No. of Villages	No. of Towns	Population	Customer No.	Wholesaler	Retailer	Mobile Retailer	Ratio
1	Pernem	251.69	26	2	77999	70199	4	26	12	1.20208333
2	Bardez	263.97	33	11	227695	204925	37	460	410	0.19861111
3	Tiswadi	213.57	26	5	160091	144081	53	459	425	0.14791667
4	Bicholim	238.8	22	4	90734	81660	4	28	5	1.57361111
5	Sattari	489.46	78	1	58613	52751	6	65	12	0.48263889
6	Ponda	292.78	28	4	149441	134496	6	62	15	1.16666667
7	Mormugao	109.1	17	3	144949	130454	42	365	250	0.17916667
8	Salcete	262.03	34	9	262035	235831	61	550	355	0.21111111
9	Quepem	318.3	36	2	74034	66630	-	8	4	3.89722222
10	Sanguem	836.8	51	2	64080	57672	6	50	156	0.22986111
11	Canacona	352.1	8	1	43997	39597	6	26	10	0.69583333
	Total	3702	359	44	1347668	1212901	225	2099	1644	0.25347222

Source: Field Survey by the Researcher, 2010

The above table depicts the talukawise distribution of fish traders, i.e., wholesalers, retailers and mobile retailers. The table also highlights the fish trader and consumer ratio in each taluka. Fish trader and consumer ratio is obtained by total number of fish consumers divided by total number of traders in particular taluka.

$$\text{Ratio} \quad \frac{\text{Total Number of Fish Consumers}}{\text{Total Number of Fish Traders}}$$

From the table no 5.5 has been observed that the study area has eleven talukas in view of size of population. The proportion of traders dealing in the sale of fish and fish products varies from one taluka to another.

Salcete taluka, Bardez, Tiswadi and Mormugao are the leading ones with regard to fish trader and consumer ratio.

Salcete taluka ranks first in the size of 2,62,035 population with density of about 895 persons per/sq. kilometer area. There are sixty one wholesalers, 550 retailers at market centre and around 355 mobile retailers. The fish trader and consumer ratio is 1:244. It implies that one trader served minimum 244 persons in respective region.

Bardez taluka ranks second in respect of the size of population i.e., 2,27,695 persons with density of 863 persons per/sq. kilometer area. There are 37 wholesalers, 460 retailers and 410 are mobile retailers engaged in fish marketing. The fish trader and consumer raio is about 1:226 (1:226 persons), that is, one trade is serving nearly 226 persons in the taluka.

Mormugao taluka ranks fourth in the size of population, that is 1,44,949 persons with density of 1328 persons per square kilometer area, which is the highest in the study area compared to any other talukas. It has a large fish market center of a variety of fish collection at Vasco. There are 42 wholesalers, 365 retailers and 250 mobile retailers. This taluka has the highest fish trader consumer ratio 1:198, which means that one trader is serving nearly 198 persons in the region.

Tiswadi taluka is an important region of the study area as it has state capital at Panaji, which is equally significant from the point of view of fish marketing, because the fish trader and consumer ratio is the highest in all talukas, i.e., 1:158 one trader serves nearly 158 persons as there are large number of wholesalers, retailers and mobile retailers. Panaji fish market is the biggest centre of fish collection and marketing.

Sanguem taluka enjoys the high fish trade consumer ratio with 1:271. The region is sparsely populated with large geographical area, that is, about 836.8 sq.km. The density of population is only about 77/sq.km. area. Due to large geographical area and sparse population in different pockets, the mobile retailing of fish is largely developed in this area. Thus there are 156 mobile retailers, 50 retailers and 6 wholesalers. Curchodem is the main market center for fish trading in the taluka.

Apart from the leading market centers, Sattari and Canacona talukas have moderate level fish trader and consumer ratio, i.e. 500-1000 consumers are served by at least one fish trader in the respective region.

Pernem, Bicholim, Ponda and Quepem talukas have lowest fish trader and consumer ratio in the study region. One fish trader is serving more than one thousand persons, i.e., Quepem taluka has just 8 retailers and 4 mobile retailers, no wholesalers. Most of the people go to the nearest biggest fish market centre either at Margao or Cuncolim to do necessary purchases. The participation of traders is very low in fish

GOA – FISH MARKET CENTERS CONSUMERS FLOW

Map-4.2

marketing in the taluka. Therefore, the ratio is the lowest compared to any other area in the study region, as one fish trader is serving 5552 persons in the taluka.

The fish trader and consumer ratio can be represented with the help of choropleth map given below.

GOA – FISH TRADER AND CONSUMER RATIO

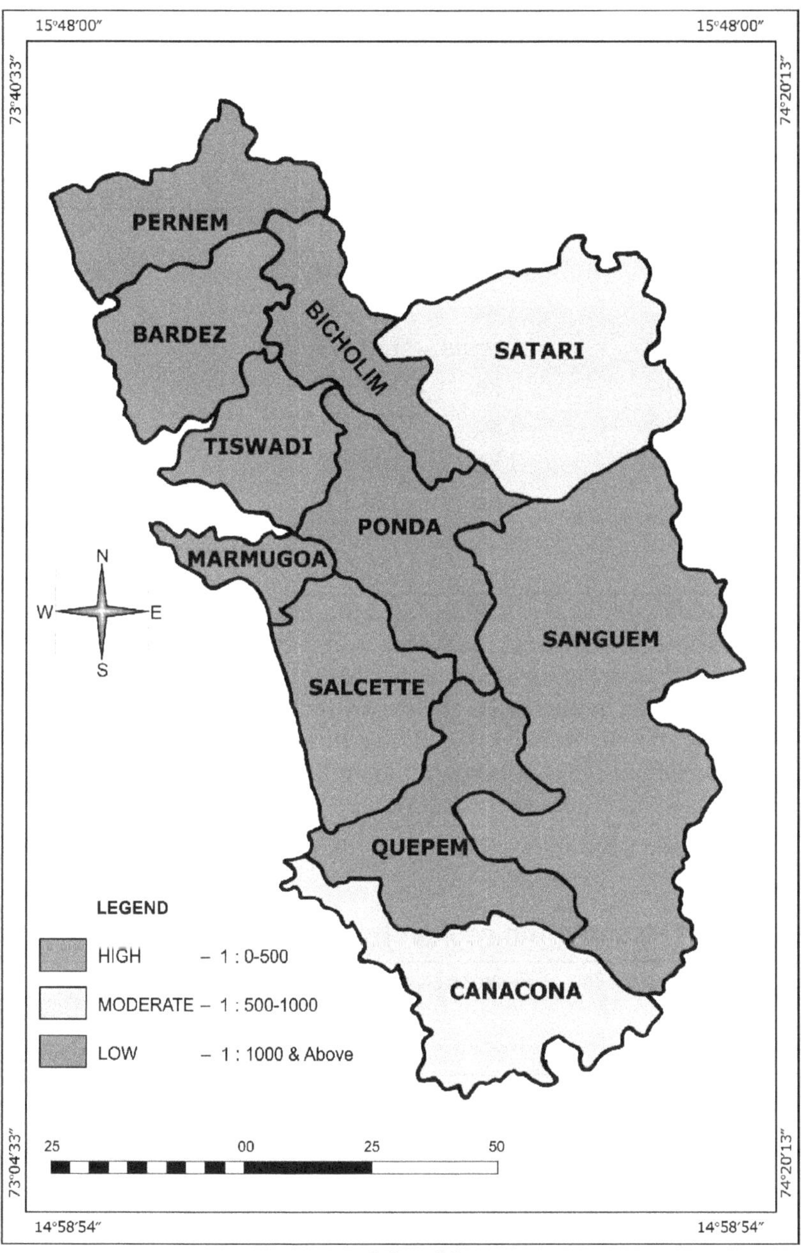

Map-4.3

The retailers selling fish in the market have to pay municipal tax at Rs. 10 per day. Those who could not dispose off the entire fish on a particular day, municipality has made provision to allow the retailers to store the fish for which the retailers have to pay Rs. 15 per day, whereas some take the fish home. Some of the wholesalers supply fish directly to hotels and restaurants.

The following table displays the economic value of selected species of fish at market centers during peak season and off season.

Table-4.6 : Value of Selected Species of Fish at Market Centers during Peak Season and Off Season

Sl. No.	Name of the Fish Commodity	Quantity Unit/Kg.	Peak Season Price in Rs. per Kg.	Off Season Price in Rs. per Kg.
1	Mackerel	8 units	100	150
2	Sharks	1 kg.	150-200	300
3	Pomfret	1 kg.	200-250	300-400
4	Lady fish	7 units	150-175	200-250
5	King fish	1 kg.	180-250	300-400
6	Tamso	1 kg.	250-300	300-350
7	Cat fish	1 kg.	100-120	150-170
8	Kurli	1 kg.	120-150	150-200
9	Milk fish	¼ kg.	50	80
10	Kampi	¼ kg.	30	50

From the above table, it can be seen that the different species of fish commodities are sold at various prices during peak season and off season period. The Directorate of Fisheries (2010) marketing officer states that they do not have direct control over the prices prevailing in the market. Prices are largely dependent upon the demand and supply situation in the market, and therefore, the prices of fish commodities are generally determined by the fishermen and brokers in the market. Thus, prices differ from peak season to off season throughout the year in Goa.

The following table shows the consumption of fish by the families in the study area.

Table-4.7 : Consumption of Fish by the Families in the Study Area

Family Size	Respondents	Percentage	Fish Consumption Daily (Kg.)	Total Consumption Weekly
4	124	62	1	6
5	40	20	1.5	9
3	20	10	<1	5
10	16	8	2	12
	200	100		

Source: Field Survey by the Researcher (2010)

Different sizes of families that consume fish in varied proportion depending upon their members and size of the income of the family. Joint families have larger quantity of consumption, whereas micro families consume low quantity of fish in their daily life.

The following tables shows the size of income of consumers and their choice of fish for consumption.

Table-4.8 : Size of Income of Consumers and their Choice of Fish for Consumption

Occupation	Respondents No.	Percentage	Monthly Income	Consumption Fish Variety
Businessmen	10	5	> 100000	Chonak, King fish, Tiger prawns
Govt. Service + Professionals	24	12	> 75000	King fish, Chonak, Pomfret
Private Service	46	23	< 25000	Prawns, Kurli, Mackerel, King fish
Labour Construction	44	22	> 15000	Local variety / Prawns
Labour Industrial	76	38	< 10000	Local variety fish i.e. Pedue, Lep, Verle, etc.
Total	200	100		

Source: Field Survey by the Researcher (2010)

From the above table it is clear that the business class of high category have high purchasing power due to their large size of income. Naturally, they prefer very good quality fresh fish for their consumption, whereas the low income group consumers have to be satisfied with the local variety of fish as they cannot afford to buy costly fish on account of low purchasing power.

The following table displays the scenario of bar and restaurants with regard to catering to the need of their customers.

Table-4.9 : Scenario of Bar and Restaurants with regard to Catering to the Need of their Customers

Sl.No.	Respondents	Percentage	Consumer Choice
1	65	65	Chonak, King fish, Crabs, Prawns
2	25	25	Mackerel, Pomfret, Shark
3	5	5	Lady fish + Local variety
4	4	4	Local variety
5	1	1	Tauso, Raus
	100		100

Source: Field Survey by the Researcher (2010)

The above table displays the restaurant and bar owners who mostly serve their customers with different variety fish in their meals. 65 percent of total serve their customers mostly with Chonak, King fish and Prawns, while one percent serve with Tauso, Raus, and the remaining serve Mackerels, Pomfret, Prawns and other local variety of fish in the study area.

It has been observed that 42.5 percent of consumers spend minimum 50 rupees daily for fish purchases, 25.5 percent spend 150 rupees, 18 percent spend 300 rupees, while 8 percent consumers spend rupees 500, and 4 percent of consumers are shelling away more than rupees 600 daily for purchase of necessary fish commodity for purpose of daily consumption.

The consumers buy fish commodities from various sources in the study area 71 percent of consumers buy fish from the nearest major market centre, 28 percent buy from mobile retailers and one percent buy from women fish vendors in the village.

The following table displays the pattern of consumption of fish commodities by different types of hotels/restaurants and bars/shacks.

Table-4.10 : Pattern of Consumption of Fish Commodities

Category	No. of Hotels	Fish – Daily Consumption
5 Star Deluxe	9	> 60 kg
5 Star	6	50 kg
4 Star	5	40 kg
3 Star	18	40 kg
2 Star	24	35 kg
1 Star	19	30 kg
Heritage	2	30 kg
Shacks/Restaurants	4000	40 kg

Source : Field Survey by the Researcher & Department of Tourism, Govt. of Goa (2010)

Retailing and Wholesaling

The following table displays the responses from retailers engaged in fish trading activities with the help of primary survey by using random sampling technique.

Table-4.11 : Retailers Collect their Fish from Different Sources

Respondents	Percentage	Source
24	24	Margao market
26	26	Mapusa market
16	16	Panaji market
6	6	No fixed wholesaler
8	8	Commurlim river
2	2	Shiroda
2	2	Siolim
2	2	Pernem
8	8	Trawler owner
4	4	Other

Source : Field Survey by the Researcher(2010)

The above table indicates that the different proportions of the retailers are collecting necessary quantity of fish from various places in the study area. 24 percent of retailers get their required fish quantity from the wholesalers in Margao fish market centre, 20 percent from Mapusa, 16 percent from Panaji, which are major fish market centres of Goa, 2 percent each from Siolim, Pernem and Shiroda, 8 percent from Cammurlim river and 8 percent from trawler owners. 6 percent of retailers are meeting their requirement of fish. There are no fixed wholesalers.

Table-4.12 : Daily Profit Margin of Retailers

Respondents	Percentage	Daily Profit Margin in Rs.
40	40	50-100
24	24	200-300
20	20	500-800
16	16	1000 & above
100	100	

Source : Field Survey by the Researcher(2010)

The above table reveals that retailers earn varied amounts of daily profits out of their fish trade in respective market centres of the study area. 40 percent of respondents earn rupees 50-100 per day, 24 percent earn rupees 200-300 per day, 20 percent retailers earn 500-800 rupees and 16 percent of retail traders can earn above rupees 1000 daily. The cashing of retailers depends largely upon the type of fish, quantity they sell and the number of customers flow in the market centres or outside in the residential areas.

Table-4.13 : Wholesaler Monthly Turnover for Fish Transaction in Rs.

Respondents	Percentage	Rs.
4	4	700000.00
10	10	500000.00
20	20	400000.00
18	18	300000.00
16	16	200000.00
32	32	150000.00
100	100	

Source : Field Survey by the Researcher(2010)

The above table displays the proportion of monthly turnover by the wholesaler with regard to fish trading in the study area. 4 percent of wholesalers do turnover rupees 7 lakh, 10 percent do the business of rupees 5 lakh in a month, whereas 32 percent of wholesalers are doing their business upto 1.5 lakh, and 20 percent of wholesalers upto 4 lakh a month.

4.7 MARKETING STRATEGIES ADAPTED IN MARKETING OF MARINE PRODUCTS

The marketing strategy is instrumental in the planning process to determine the effective measures in order to overcome the challenges identified in the fish marketing system. Maintaining high quality food and fish should be propagated as a strategy to stay ahead of other competing countries in the world market Anjani Kumar (2003).

1. **Quality standards**:- form a system of classification which helps to make the market more transparent by indicating certain characteristics of a product marketing standards, reduce transportation cost and allow trade to develop without physical attraction to product.
2. **Product information**:- information of the fish commodity should be made familiar to the customer, i.e., species and variety of fish, which leads to good demand for consumption.
3. **Price**:- pricing of fish and fish products should be on reasonable terms, prices have to be kept low/medium during season, otherwise common people cannot afford to buy such high priced commodity, which ultimately affects consumption pattern of the people.
4. **Promotion**:- encouraging retailer, sub agents by giving them good margin on the sale of variety of products of fish, which is essential in marketing of fish.
5. **Integrated catch and supply**:- the necessary efforts should be made to increase the catch and supply of variety of fish products in respective market centers to ease out the pressure on demand-high price so that the consumers feel good to buy and consume fish at reasonable prices.
6. **Performance**:- providing quality service is the important consideration in fish marketing, customers should be given fresh commodities rather than being cheated.
7. **Infrastructure**:- providing good market centers-platforms, cold storage facilities, sanitation facilities, is very important for fish retailers and wholesalers to increase the sales of fish and fish products. These measures will definitely encourage many people towards fish trade.
8. **Finance**:- easy availability of finance to fish traders to take up marketing activity with great zeal should be ensured. This helps in overcoming various problems of fishermen community and fish traders, i.e., the State Bank of India, Central Bank, Canara Bank, etc. and other government agency must encourage fish trade community for their better survival in the study area.

Hence the hypothesis that the local markets determine the economic status of the fishermen community, whereas the global market determine the economy of Goa has been tested with literature and confirmed the hypothesis.

Chapter-5

PLANNING AND STRATEGIES

5.1 INTRODUCTION

There are six million people in India who are engaged in the catching, marketing and processing of fish. Unlike most other jobs where people work for someone else, most of the fisher folk own their means of production (nets, boats, etc.) which thus gives them full control of their economic production (2009-10) Directorate of Fisheries.

In Goa, there are about 36,891 people actively involved in fishing activity. The fishermen are courageous, and are ready to face all the dangers of the open seas, at great risk to their lives, to provide cheap and protein rich diet for the Goan masses.

They have a deep self respect for their jobs, and are not lazy as some call them. The men go out in the crafts very early in the morning to catch fish, sometimes even at midnight. When they return home, the women in the family take care of marketing the fish. During day the fishermen spend their hours mending their nets and crafts. They have a high sense of dignity of labor, and one can consider their profession as a tremendous service to the society, unlike other profession. Though lacking in formal school education, yet they understand and have a genuine concern for the ecology of the seas that sustain them in all respect.

5.2 THE GOAN FISHERMEN'S STRUGGLE

The traditional fishermen of Goa have been struggling for more that 15 years to prevent the ecological disaster of the Goan inshore waters, caused by trawlers and purseiners. Earlier fertilizer factory along the Salcete coast began dumping poisonous effluents into the sea. This pollution killed shoals of fish, the fishermen, along with the concerned villagers, students, teachers and other citizens, staged various protests against the factory demanding that it be closed, until it set up an effluent treatment plant. Then the voice of the people triumphed. This success gave a boost to the fishermen, and it became evident to them that, if they are united, they could solve the problem of trawlers and purseiners encroaching on their traditional fishing waters. With the active support of Mr. Mathany Saldanha, they came together to form the "Goenchea Ramponkarancho Ekvott" (The All Goa Fishermen's Union). They demanded the immediate ban of trawling and purseining in inshore waters and warned that if this demand was ignored, the Goans would export the produce, thus depriving the local population of one more source of fish.

The fisherwoman is a very familiar sight in Goa. The practice in most Goan fisher-folk families the husband to go out fishing and for the wife to take the responsibility of marketing the fishes.

The market has been exposed to considerable commercialization with more and more fish production being consumed by the export and tourism sector. Traditional preservation methods are being replaced by cold storage facilities. Commercial interests have acquired a crucial monopoly over net manufacture-mechanization in this area has resulted in the displacement of people. Fish fertilizers having become unavailable, there is increasing dependence on commercially produced fertilizers.

The implications of all these changes are far reaching and affect every facet of life in Goa. In particular, the ecological implications of fisheries development are profoundly distributing. There seems to be, however, no adequate response on the part of the administration. The powers-that-be seem securely locked in the imported conundrums of the development process.

5.3 PROBLEMS OF FISHERIES AND FISHERMEN

Fishing is an important occupation of the people of Goa, as it provides a large number of employment opportunities for the people of various talukas in Goa. According to recent information, there is a decline in the importance of fishing activities due to several problems faced by the fishing industry. These problems can be divided into three main categories as discussed in this text.

1. Environmental Problem,
2. Socio Economic Problems, and
3. Personal Problems.

1. Environmental Problem

(a) Gusty Winds

Gusty winds blowing at sea prevent trawler owners as well as traditional fishermen using motor boats from venturing into the sea. Fishermen do not venture into the sea as a precautionary measure due to high intensity winds blowing at sea. No trawler owners from any of the fishing jetties in the state Malim, Vasco, Cutbona, Betul, Chapora risk venturing into the sea. All the trawlers, though ready for fishing, lie berthed at jetties due to inclement weather. Due to bad weather, the fishermen who make all arrangement for moving into the sea to catch fish, decide not to take risk and venture into the sea till sea conditions stabilize, where there is a probability of losing the catch.

(b) Pollution

Shrimp farmers often overstock ponds with shrimp larvae with a view to producing high yields. Unfortunately, the facial pellets and left over nutrients, during the feeding of shrimp in ponds, has increased pollution and diseases in the wild fish and other marine fauna. Today, around 20 percent of the fish not only increase morbidity but also lower the nutrient and economic value of the catch.

(c) Monsoon Season

The monsoon season also coincides with the peak breeding season for most fish on the west coast of India. Due to heavy monsoon showers, the lakes, ponds, are also over flown. As a result, the "prized catch", including solar prawns diminish.

(d) Accidental Problems

The accidental problems are also main hurdles not only for the fishermen but also for natural environment. There was an accidental oil spilling off the Aguada coast, which was later drifted along the South West coast of Goa. Then the state administration alerted the owners of the mechanized fishing vessels and the traditional fishermen not to catch fish till fish stops smelling of petroleum products. The oil spill also adds to the problem by causing pollution and destroying the flora and fauna at the sea.

(e) Other Environmental Problems

There has been gradual decline in the supply of fish due to limited catch as a result of several environmental and other problems. Survey revealed that today there is a considerable increase in the levels of fish morbidity and mortality due to environmental degradation pollution caused by the toxic effluents and discharges from land and oil spillage from the barges, fishing vessels, ships, etc. in the sea water.

Major jetties as well as others have excess trawlers. Therefore, the Directorate of fisheries has to take a decision not to issue any more new licenses, except when trawlers have to be replaced due to sinkage and irreparable damages, etc.

(f) Disturbance during Fish Breeding Season

Fish breeding season is usually in the month of June and July. The fish must not be disturbed during the fish breeding season, but it has been observed that despite the ban few boats and trawlers are venturing into the sea, and this has adversely affected the spawning activity.

(g) Lack of Appropriate Nets

It is not just the fish that counts, but also the size of the fishermen's nets, their hooks and the size of their boats. The government had imposed certain restrictions on the size of nets of a minimum mesh size of 35 mm to prevent catching of baby fish. Despite such restrictions, it has been observed that some boats even use nets of mesh sizes of 15 mm due to which the baby fish is trapped, and hence the whole life cycle of the fish is adversely affected.

2. Social and Economic Problems

Social and economic problems have been identified in the present study and are given below:

a) Fishing craft requires adequate harboring. Landing and other shore facilities. At present, there are no separate harbors or jetties for the sole use of the fishing industry in the Bardez taluka.

b) There are no adequate servicing facilities like workshop for repairing crafts, since mechanized vessels require servicing facilities

c) Though there is a need for more and more ice for fish preservation, there are no efforts made for cold storage plant establishment for handling fish production.

d) The fish, which is found in abundance on the Goan coast, if cured can fetch a good price. In the absence of curing facilities at important fishing villages, fishermen do not get economic prices.

3. Personal Problems

a) The standard of living is a prime indicator of modernization. Due to modernization the youth from Kharvi community in Goa are slowly shying away from the prosperous and lucrative trade, leaving the marine resources of our coastal ecosystem to be exploited and harnessed by enterprising migrants. Today, the demography of the coastal Goa during fishing season has changed rapidly. Though licenses and permits are given to nearly 1,000 trawlers, most of them employ migrant labor force.

b) Challenging activity: Fishing is a challenging activity as men at sea have to fight a lot with both the waves and the winds, especially during high tides as the sea is rough, and there is a possibility of drowning due to cyclonic winds blowing. So the fishermen do not take the initiative of venturing into the sea, when the sea is rough.

c) Disturbance of peace and increase in crime: Rough estimates show that Goan manpower requirements on the fishing vessels works out around 15,000 personnel, of which hardly 2,000 come from the local market forcing the boat owners to go in search of the labour across the borders. This migrant influx in Goa has not only changed the ethnic demography but has disturbed peace and increased crime.

d) Implementation of various schemes: There is lack of various state and central government schemes meant for the welfare and betterment of fishermen, especially in the rural areas, and hence they lack zeal and motivation.

e) Old, traditional methods of catching fish: Many fishermen use the old traditional methods of catching fish, especially because they are not well aware of the subsidies provided by the government or because they are not aware of the new methods of catching fish. Some of them are fully aware that due to their economic poverty, they could not afford to switch over to the mechanized system.

4. Marketing Observations

1) Fisherman catch all kinds of fish in their nets, they do not keep all of them, but throw a number into the sea, which then float to the seashore. Such kinds of fish cannot be marketed due to their spoilage and tiny size.

2) There is an increasing demand for fish and fish products all over Goa owing to increasing density of population. The growing tourism, industrialization and other economic sectors influenced great demand for fishery products and raw fish.

3) Shri. Pokle, a marketing officer states, "There is no direct involvement of Government agency towards marketing of fish products and there is no direct control over the prices prevailing in the fish markets of towns / cities / villages throughout the year in Goa. The price will be decided by the traders, i.e., wholesalers and brokers depending on the demand and supply situation with regard to fish and fishery products". The Directorate of Fisheries, Government of Goa (2010).

5.4 NECESSARY MEASURES TO SOLVE VARIOUS PROBLEMS

It is time for the Goans to respect nature too by endeavoring to extend the fishing ban period from June 15th to August 15th to maintain the growth of fisheries, and to prevent the fish coming to breed in the mangroves from getting disturbed and destroyed. The dumping of garbage in the mangroves that will destroy fish nurseries should be strictly banned. Fish breeding season is usually in the month of June and July.

Uniform fishing ban along the entire west coast of India from Gujarat to Kerala must be imposed by the Government of India

Do not allow hospital waste, batteries and urban commercial debris to be dumped in our creeks, that will allow mangroves to develop and act as nurseries of fish and be a gene pool for biodiversity.

To protect the magnificent catch, an animal welfare organization has to start awareness campaign among the fishermen community about their vital role in the marine ecosystem. Effective implementation of ban in the territorial waters on fishing and prevention of its violation is necessary, especially during the monsoon season as the monsoon season is the spawning season of demersal fish and shrimps.

If we do not take into consideration the above points there will be an increase in fish diseases which will adversely affect holistic growth of fisheries and development of Goa. Unless technically qualified persons are allowed to manage the fishery resources of Goa, the state would be reeling towards self destruction from where it would be no coming back.

The Role of the Government of Goa

Directorate of fisheries provides infrastructure facilities, deals with the designing of fishing vessels, fishing canoes, fishing crafts, fishing nets, etc., besides providing financial assistance to the needy fishermen by way of subsidy on purchase of out Board Motors for fishing crafts. It also provides for fishing trawlers purchased through financial institutions.

The Directorate of Fisheries extension wing, consisting of extension officers and Gramsevaks attached to community development blocks of Tiswadi, Bardez, Salcete, Morrmugao, Ponda, Pernem and Canacona to take up developmental activities.

The Directorate is running a fisherman training center and one estuarine fish farm, fresh water fish farm at Anjunem, a sub office at Colva, ice factory and cold storage facilities at Patto-Panaji and Chapora and Pilot Prawn Hatchery at Benaulim are operated under Brackish Water Fish Farmers Development Agency-Goa.

1. Fishermen Training Course: The fishermen training center offers a course meant for training in fishing craft and gear, fundamentals of navigation and seamanship, maintenance and operation of marine diesel engine, weaving and mending of modern nets. Candidates, who have passed std. VIII and are between the age group of 18-25 years, are eligible for the course. Preference is given to scheduled caste and scheduled tribes candidates as well as young persons having experience at sea.
2. Government Support: The government has made available in all over Goa fishing jetties, work shop stalls, offices for fishing societies, canteen facilities, HHD pumps (diesel), water supply, parking space for vehicles, net mending sheds, workshops, sulabh toilets, etc.
3. Employment Opportunities: Govt. has declared that it would ensure that 80% of jobs would be reserved for Goan Youth in Goan industries, with preference being given to scheduled caste and scheduled tribes.

5.5 MANAGEMENT OF FISHERIES

For sustainable development of fisheries sector and professional management of resources, it is necessary to evolve a state policy on fisheries. This has become necessary due to reports on fish depletion on account of once exploitation, degradation of breeding grounds, fishing during ban periods, discarding of low value fish, etc., and threat to ecology due to fast growing aquaculture. With increase in the fishing fleet, new avenues for export of certain fish and sea food and encouragement by the Government of India to exploit deeper waters in E.E.Z. it has become imperative to provide necessary infrastructure and integrated shore facilities, like jetties along with ice factories, cold storages and other shore amenities.

The broad objectives envisaged for development of fisheries sector in the state are as under:

1. Evolving a state policy on fisheries;
2. Establishment of fish processing industrial estate;

3. Construction and mechanization of fishing crafts;
4. Increasing fish production by utilization of the available natural resources;
5. To ameliorate the socio-economic conditions of the fishermen, who belong to the weaker sections of the society;
6. Development of storage and marketing infrastructure;
7. Development of fresh water fish culture to increase inland fish production;
8. Development of eco-friendly brackish water aquaculture.

To achieve the above objectives, the following schemes are proposed:

1. Evolving State Policy on Fisheries

The schemes envisage evolving a state policy on fisheries for sustainable development and professional management of resources. Fish, being a natural resource, needs to be managed professionally for continuous biological replenishment of reserves and to ensure regular economic benefits to the fishing community, and development of economy of the state. There have been reports regarding fish depletion on account of destruction of breeding grounds due to over exploitation, fishing during breeding season, wastage due to discarding of low value fish, etc. Fast growing aquaculture activities in the state may lead to destruction of mangroves, conversion of agricultural land, wet land, etc., in spite of clear guidelines. It is in this context that the Government intends to evolve a state policy on fisheries to rejuvenate the fisheries sector in the state. This would involve marine survey of aquatic resources, assessment of environmental impact, evaluation of demand, export potential, employment opportunities, assessment of infrastructure requirement, etc.

2. Establishment of Fish Processing Industrial Estate

The scheme aims at encouraging entrepreneurs to establish units for commercial production of value added products from fish. It is observed that low value fish is generally discarded by the trawlers and fishermen, which can be used for production of value added food products, like protein concentrates, water, sausages, soup, cutlets, etc. For this purpose, it is proposed to set up a separate fish processing Industrial Estate in the state with the necessary infrastructure facilities.

3. Development of Infrastructure for Fisheries

The scheme envisages development of various infrastructures, like landing and berthing facilities, construction of roads; net mending sheds and auction halls, provisions of water supply, HSD outlets, illumination, transport facilities, development of fishing harbours, etc. The work on the retaining wall for the protection of jetty is in progress and the work on the administrative building at Cutbona will be taken up. So also the work of overhead tank at Cutbona is in progress. Construction of net mending sheds at Madkai, Assolna and Canacona will be taken up during the year.

4. Mechanization/Motorization of Fishing Crafts

The scheme envisages providing financial assistance to traditional fishermen for construction of fishing crafts and purchase of outboard motors(OBM) for the crafts to enable them to reach distant fishing grounds for better fish catch. For this purpose the following programmes are proposed:

(a) For construction of fishing craft either of wooden or F.F.Plastic, subsidy to the extent of 50% of the cost limited to Rs.20000 per craft, will be provided. During the year 2003-2004, 25 traditional fishermen were provided subsidy and 4 fishermen during the year 2004-2005. It has been proposed to provide subsidy to 20 fishermen among 2010-11.

(b) For purchase of OBM, subsidy will be to the extent of 50% of the cost limited to Rs.10000 per OBM. This is a centrally sponsored programme and the cost is met on 50:50 basis between the State and Central Governments. During the year 2003-2004, 22 fishermen were provided subsidy. Subsidy also provided to 25 fishermen during the year 2004-2005. It has been proposed to provide subsidy to 30 fishermen during 2010-11.

(c) It is proposed to reimburse excise duty charged on diesel oil to mechanized vessels for fishing beyond 5kms from the coast with vessels below 20 metres registered with the Directorate of Fisheries, under the marine fishing regulation act. This is a centrally sponsored programme on 20:80 basis between the State and Central Governments. It is proposed to cover 600 beneficiaries during the year 2010-11.

(d) It is also proposed to provide further relief to the fishing community by way of subsidy to cover the fluctuation of rates in diesel prices. Any average increase over the ceiling price of diesel during the year will be compensated by way of subsidy subject to a maximum of Rs. 75000 - 100000.

(e) With a view to giving relief to the traditional fishermen, who are compelled to buy kerosene at high prices from the open market, a special quota of kerosene will be requisitioned from the Government of India and the same will be made available to the traditional fishermen at the depots from August 15, 2004.

(f) Supply of insulated boxes under this scheme whereby a fisherman/fish vendor is entitled to subsidy for purchase of insulated boxes in order to preserve the fish in hygienic condition for marketing purpose. The applicant engaged in selling/marketing of fish will be eligible for one box of 25 litres and 50 litres capacity each and will be entitled to a subsidy to the extent of 75% of the cost limited to Rs. 1500/-. The fishermen actively engaged in fishing at sea will be eligible for the boxes of 50 liters and 100 liters capacity each and subsidy to the extent of 75% of the cost limited to Rs. 3000/-.

(g) A subsidy based on the Goa Value Added Tax (VAT) on H.S.D. oil consumed by fishing vessels scheme 2006 is proposed to provide relief to the operators of the fishing vessels to overcome financial losses suffered by them on account of the ever increasing cost of the fuel, the H.S.D. oil, so as to enable them to sustain themselves. An amount equivalent to the VAT tax will be reimbursed to the fishermen, limited to Rs. 6/- per liter on the diesel consumed by the fishermen on a total quota of 16000 kilogram.

5. Providing Storage and Marketing Infrastructure

The scheme envisages providing financial assistance to local bodies and to fishermen for construction of fish markets, purchase of insulated boxes and cycles and auto-rickshaws fitted with insulated boxes and deep freezers. For this purpose, the following programmes are proposed:

(a) For construction of fish markets, subsidy to the extent of 50% of the cost of construction limited to Rs. 3.00 lakh will be provided to Municipalities and Village Panchayats. It is proposed to provide subsidy to 2 Village Panchayats/ Municipalities during the year 2004-2005.

(b) For the purchase of insulated boxes, it is proposed to provide subsidy to the extent of 90% of the cost to traditional fisher men. During the year 2004-2005, it is proposed to provide subsidy to 400 fishermen.

(c) Provision of subsidy to the extent of 50% of the cost limited to Rs.1500 for purchase of cycles and Rs.25000 for auto-rickshaws fitted with insulated boxes.

6. Development of Freshwater Fish Culture

The scheme envisages developing vast area covered under freshwater bodies for production of fast growing freshwater fish varieties like Rohu, Katla, Mrigal, etc. This would increase overall fish production in the state, and would also ensure availability of fish during the monsoon season. For this purpose, the following programmes are proposed:

(a) To provide subsidy to entrepreneurs for construction/renovation of perennial freshwater tanks to attain minimum water depth of 5 feet before monsoons. Subsidy will be provided to the extent of 25% of the cost limited to Rs. 40000 per ha. During the annual plan 2003-2004, four beneficiaries were covered. It is proposed to cover ten beneficiaries during the annual plan 2004-2005.

(b) In order to provide demonstration and training in freshwater fish culture to interested entrepreneur, it is proposed to establish an Integrated Pilot Freshwater Fish Farm at Kerim, in Sattari during 2004-2005.

(c) Financial assistance in the form of loan will be provided by Nationalized Banks and other financial institutions.

(d) Subsidy to cage and pen culture is provided to the extent of 50% of the total cost of the cage or pen (5000 sq.mtrs.area) limited to Rs. 20,000/- in case of pen and Rs. 30,000/- in case of cage and Rs. 5000/- subsidy for feed and seed. This will be a one time subsidy.

(e) The entrepreneurs should have clear ownership title of land/water body covering the tank or should have taken the land/water body on lease for at least 3 years at the time of application.

(f) Goa, being a tourist destination, it is proposed to set up a fresh water fish Aquarium at Kerim in Stattari Taluka during 2004-2005 which would also be useful for educational and research activities.

7. Development of Brackish Water Fisheries

The scheme envisages developing brackish water resources, which would generate supplementary income to marginal fish farmers. The following programmes are proposed under the scheme:

(a) To provide subsidy through the Brackish-water Fish-farmers Development Agency (BFDA) for development of fish farms. Subsidy of Rs. 30000 per ha is provided for a maximum of 5 ha of which Rs. 20000 is for capital investment and Rs. 10000 is towards input cost. During the year 2003-2004, 15 fish farmers were assisted. It is proposed to assist 10 farmers during 2004-2005. This is a centrally sponsored programme and the expenditure is met on 50:50 basis between the State and central Governments.

(b) To produce 10 million prawn seed during 2004-2005 at the Pilot Prawn Seed Hatchery at Benaulim in Salcete Taluka operated by the BFDA and the same will be supplied to farmers at a reasonable price for prawn cultivation. During the year 2003-2004, 5 million prawn seeds were produced and sold out to the farmers, generating revenue of Rs. 22.00 lakh.

(c) To organize demonstration cum training programmes to entrepreneurs at the departmental Fish farms at Dauji, Old Goa, regarding application of new technology in prawn and fish farming so as to maximize per unit output. It is proposed to impart training to 100 entrepreneurs during 2004-2005.

(d) Towards ensuring sustainable eco-friendly aquaculture in the State, it is proposed to set up a Diagnostic Laboratory in the State in collaboration with ICAR/NIO

8. Direction and Administration

The scheme envisages supervision, control and monitoring of fisheries sector development in the state. For this purpose, the following activities are proposed:

(a) To computerize all the activities of the Department and to develop a statistical database;

(b) To provide training to all the officials of the Department as regards latest developments in fisheries sector so as to increase their efficiency and productivity;

(c) To take up minor works related to upkeep and maintenance of the building.

9. Enforcement and protection of Reserved Fishing areas along the Goa Coast

The following activities are proposed under this scheme:

(a) To strictly enforce fishing ban period and also to guard the restricted zone of waters along the Goa coast within a specified area for mechanized fishing in order to protect the interests of traditional fishermen and also to ensure conservation of fishery resources in keeping with the provisions of marine Fishing Regulation Act, 1781.

(b) To facilitate enforcement activities, a patrol boat has been constructed and commissioned during the year 2003-2004.

10. Financial Assistance to Fishermen for purchase of Fishery Requisites

The scheme envisages providing subsidy to fishermen for purchase of fishery requisites, like gill nets, monofilament twine, nylon twine, sinkers with accessories and Rs. 5000 for nylon webbing accessories, floats, sinkers, ropes, etc. During the year 2003-2004, 28 fishermen were provided subsidy. An outlay of Rs. 3.00 lakh is proposed in the annual plan 2004-2005 for providing subsidy to 50 fishermen.

11. Education and Training

The scheme envisages providing education and training on technological advancements in fisheries sector to fisher youth as well as to officials of the Department. For this purpose, the following programmes are proposed:

(a) To impart training to fisher youth in the operation of mechanized vessels and other modern techniques in fisheries at the Departmental Training Centre. Some deserving fishermen are also sent for advanced operative courses at the Central Institute of Fisheries and Nautical Engineering, Cochin. During the year 2003-2004, 25 fisher youth were imparted training. It is proposed to maintain the same target for the annual plan 2004-2005.

(b) To depute the departmental officials for training at the ICAR Mumbai, CMFRI Cochin, and other institutions in the country. During the year 2004-2005, it is proposed to depute 2 officials for training.

12. General Welfare Scheme for Fishermen

This new scheme has been introduced in the annual plan 2004-2005, under which the following components will be covered viz. (a) social security, (b) general insurance, (c) welfare relief, (d) saving cum relief fund, (e) development of fishermen village:

(a) Social Security scheme for fishermen is a scheme aimed at the deprived and most vulnerable section of the fishermen community in the age group of 50 to 60 years only. Under this scheme, it is proposed to give financial assistance of Rs. 500/- per month to each beneficiary, which will eventually pass on to the spouse and children on the expiry of the beneficiary. There are around 5000 fishermen families existing in Goa. Initially, it is proposed to cover 500 families during the year 2004-2005. It is expected to cover all families by the end of the 10th Five Year Plan 2002-2007.

(b) General Insurance Scheme for fishermen envisages providing insurance cover to all active fishermen in the age group of 18 to 60 years, living below the Poverty Line. The fishermen, marginally above the poverty line, will also be covered under the scheme. The benefit will be paid in all accidental deaths, including death occurring on the high seas due to natural calamities. Benefits on the event of death by accident in partial/total/permanent disability are as under:

i. On death due to accident Rs. 50,000

ii. Permanent/total disability due to accident Rs. 50,000

iii. Partial disability due to accident Rs. 25,000

Besides the above, in the event of natural death of the beneficiary, a sum of Rs. 20,000 will be paid to the nominee. It is proposed to provide insurance cover to 2000 active fishermen.

(c) Welfare Relief fund for fishermen aims at providing immediate relief to fishermen who are victims of unforeseen natural calamities and accidents, since at present there are no means of providing any other immediate relief to the helpless fishermen. Immediate financial assistance will also be rendered to fishermen in the event of extensive damages or robberies of fishing nets operated by them. In order to avoid undue delay in completing the necessary formalities and to provide immediate assistance to the fishermen in distress, a committee under the Chairmanship of Minister for Fisheries will be constituted, under which, an emergency financial assistance upto Rs. 1000 will be granted.

(d) Saving cum Relief fund scheme under which scheme Rs. 100 shall be collected from all eligible fishermen for a period of 8 months in a year and a total of Rs. 1200 thus collected will be matched with 50% contribution from the State and Central Government separately (i.e. Rs. 600 each). The total sum of Rs. 2400 thus collected will be distributed during the four lean months to the beneficiaries in four equal monthly installments. The interest accrued will be disbursed with the 4th Installment.

(e) Development of fishermen village under which the eligible fishermen would be provided with basic civic amenities, like housing, drinking water and a common place for recreation. As regards the allotment of houses under this scheme, it will be ensured that the beneficiary is an active fisherman. Preference will be given to the landless fishermen below poverty line. Fishermen owning land or katcha structure will also be considered for allotment of houses. Under this scheme,

i) A fishermen village will consist of not less than 10 houses and the plinth area and cost of construction of house would be limited to 35 sq.mts. and Rs. 40,000/-. The Central and State Governments will share the cost of the development equally.

ii) Financial assistance for providing drinking water to backward fishermen area. Scheme will have the provision for drinking water facilities to backward fishermen area, where there are no drinking water facilities.

iii) The fishermen area shall have more than 10 houses in number. The charges of water supplied through public tap shall be borne by the Government.

iv) The charges for water supplied to the individual houses of the fishermen shall be borne by himself individually. The cost of providing water facility shall not exceed Rs. 30,000/-

v) Financial assistance for construction of Community Hall/Work shed for recreation and common working place in a Fishermen Village having atleast 75 houses will be provided by the Government.

vi) The hall/shed shall be constructed on an area not exceeding 200 sq.mts. and shall have 2 toilets separately, one for ladies and one for gents.

vii) The area required for the Community Hall shall be provided by the Government, gifted by any landlord, donated by the Communidade or purchased by the Panchayat and donated to the Government.

viii) The Community Hall shall be managed by the Panchayat or Fishermen Association and day to day maintenance of the Hall shall be done by the Panchayat/Fishermen Association in whose jurisdiction the Hall is situated.

ix) The Panchayat shall charge nominal fee for the programme of individual fishermen, if arranged in the Hall. However, public programs of the fishermen shall be free from the imposition of any charges.

x) The amount of charges shall be on the basis of no loss, no profit.

13. Construction of Fish Markets

Financial assistance in the form of subsidy shall be given to local bodies such as Municipalities and Village Panchayats for construction of fish markets as per the following pattern:

Sl. No.	Panchayat/Municipality of Having Annual Income	Assistance of Quantum
1	Upto 2 lakhs	100% subsidy limited to Rs. 5.00 lakhs
2	Above 2 lakhs	75% subsidy limited to Rs. 5.00 lakhs

Documents required:

1. Panchayat/Municipality resolution,
2. Land documents complete in all respects,
3. Estimates duly approved by P.W.D.,
4. Site plan duly approved by respective authority.

Fisheries Co-operatives in Goa

As early as 1964, the Department of Fisheries in collaboration with the Registrar of Co-operative Societies, motivated the fishermen to organize primary fisheries co-operatives. There was a good response. Altogether 10 primary fisheries co-operatives were formed in important fishing villages in the then Goan territory. However, over the years, the major co-operatives became either dormant or got de-functioned. The only society which continues to function is the Akhil Gomantak Sarkari Sahakari Saunstha Limited, Goa with their regular office at Panaji which was established in 1970, and has now 1500 members. This is a Society of the fishing stake operators, having their fishing activity in the inland area in the rivers and estuaries.

During the last ten years, 2 more fisheries co-operatives have been formed. They are as follows:

Sl.No.	Name of Society	No. of Members
1	Mandovi Fishermen Marketing Co-operative Society Ltd. Panjim	250
2	The Chapora Boat Owners Fisheries Co-operative Society Ltd. Chapora	47

The main activity as on date of the above two active societies have been operating high speed outlets. They also coordinate for sale of their catch to the fish merchants and look after the common interests of the fishermen members.

Among the two Co-operative Societies, the Mandovi Fishermen Co-operative Society has been the most active society which is presently operating from Malim fish landing centres. The Society has been assuming major role in maintaining their fish carrying centre and is operating a store of fisheries requisites for the benefit of its members.

The members of the two co-operative societies, who purchased HSD oil from the outlets operated by the said co-operative societies, are entitled to exemption of excise duty under the centrally sponsored scheme.

It is hoped that the above co-operative societies will assume more responsibilities in undertaking functions beneficial to the member operators and in improving fishing industry as a whole.

The Dept. of Fisheries during the year 1995-96 allotted walkie-talkie sets to the members sponsored by the two fisheries co-operative societies, and these sponsored by the 2 associations, one representing traditional fishermen and the other representing mechanized boat owners.

Table-5.1 : Number of Fisheries Co-op. Society Existing in Goa

Sl. No.	Name of the Society	No. of Members Registered
1	Akhil Gomantak Harkari Saunstha Ltd., Naik Villa, Panaji	1500
2	Mandovi Fishermen Marketing Co-op Society Ltd., Betim	250
3	Xapora Boat Owners Fisheries Co-op Society, Xapora	47
4	Cutbona Fisheries Co-op Society Ltd. Salcete	230
5	Riosal Fisheries Co-op. Society Ltd. Salcete	51
6	South Goa Mechanized Boat Owners Co-op. and Marketing Society, Cutbona	78
7	Zuari Fishermen Marketing Co-op. Society, Vasco	62
8	Vasco Fishing Boat Owners Marketing Co-op. Society, Vasco	128

Source: Directorate of Fisheries, Govt. of Goa, (2010)

Thus appropriate measures are taken by the Directorate of Fisheries, Government of Goa. Brackish Water Fish Development Agency and other leading organization to ensure sustainable growth of fishery resource of Goa in general and the welfare of fishermen community in particular as well as fish traders in the study area.

Hence, the hypothesis that the planning strategy for the fishery activities are not only for prosperity of the fishermen community, but also related to the development of the region has been tested with literature and confirmed the hypothesis.

Chapter-6

CONCLUSION, FINDINGS AND SUGGESTIONS

6.1 CONCLUSION

Geography is a multidisciplinary subject, and is gaining much importance. It has attracted not only by the academicians and planners, but also administrators, which provides enough scope to take up research from the micro level to the macro level of any geographical regions in the world. As a result, such a kind of research output helped the planners as well as administrators to a great extent for the development of any region. The latest technology is also responsible for fast growing knowledge in the field of subject. Hence, the discipline plays a significant role in the modern world.

Fish has become an important source of food for many centuries. Even today it is also the prime source of food substances in many countries. Earlier fish capturing was practiced by men with traditional methods and by applying their skills. Today the modern means of technology are at their disposal. As a result, the modernization has been adopted along with traditional methods. The mechanization adopted in marine fishing and traditional methods are practiced not only in aquaculture but also in marine fishing activities. The high level mechanization is found in the technically advanced countries, and vice versa.

Fishery activity has its own historical background, and now-a-days it is also one of the main dominant determinants from the economic point of view in the coastal area, in particular, and source of water resource in general.

Fisheries and agricultural farming have evolved rather parallel in the history of human civilization. Interest in fish eating dates back to the dawn of the history. It is believed that hunting of fish was not uncommon in prehistoric times. At the developing sites near a river or lake of cave dwellers of the late Old Stone Age (40,000 BC) heaps of refuse of shellfish and sea fish have been found. The great distance by which these sites were separated from the sea points to some primitive way of fish preservation (like sun drying and smoke drying over the fire wood) practiced by these ancient dwellers to keep the food in edible of salmon smoking practices. Bronze Age (3500 BC) was the time when salting of fish started. Trading in dried fish, of course, was in vogue with ancient civilizations of Egypt, Mesopotamia and Indus Valley.

The role of fishing has changed to a considerable level over the number of years. In the primitive time, fishing was mainly for consumption, but today in the modern world fishing is vibrant economic activity. The traditional method of fishing, too, has been replaced by modern scientific methods of fishing. Fishing is an activity, which is carried on in all the parts of the worlds, and the use of fish, too is varied.

The fishing activity provides employment either directly or indirectly to many people of Goa, and also contributes significantly to the total economic growth of the state. It gives boost to the canning industries, ice factories, boat making, basket making, etc. A couple of oil extraction as well as fish meals manufacturing units are also set up in the state. The fisheries also help in earning foreign exchange. Because of all these reasons, the fishing sector is developing rapidly.

More than 200 million people in the world are involved in fishery activities which contribute 3% of working population.

India's plentiful bounty of nature and biodiversity in regard to fish germplasm resources are amazing. The country is endowed with abundant inland and marine germplasm resources. India is the third largest producer of fish next only to China and Peru, and it ranks second in the aquaculture production farmed inland fish in the world.

In earlier days, the term marketing of fish meant buying and selling of fish at the leading centers. After the Second World War, marketing of fish has taken a new role in business activity.

Marketing Geography is a recent origin of the late 1950's. It has attracted the Geographers and planners, and accordingly moved in the research activities. Marketing Geographers are interested in trading areas, which are complex in real manner. Therefore, they seek to provide scientific knowledge and scientific assistance to business men in evaluating market area potentials. Marketing is an integral part of economic geography. There is no doubt that marketing is a part of man's economic activity.

Goa is one of the tiny states of India. It has ecstatic beauty that attracts the tourists from all over the country as well as world, and therefore, popularly known as a tourist state. It is well endowed with natural resources, i.e., minerals, forests, fisheries as well as agriculture. The mining activities are predominant, and also a key factor for industrialization. Goa is also known for popular drink cashew fenny. As a result, these aspects are determining the economy of Goa. Because there was a colonial rule by the Portuguese for a period of 450 years in Goa, still the impact of Portuguese culture can be seen all over Goa. The movement of people from other states brought the diversity in the cultural environment of Goa. Therefore, an attempt has been made to understand the geographical personality of the study area, in terms of its physiography, geology, climate, drainage, natural vegetation and soil, etc. in the form of physical characteristics.

The demographic characteristics are discussed in length. The settlement characteristics are the main considerations in the present study area, taking the population into account. The economic characteristics which form sound base for the economy of Goa. The transportation is the prime indicator not only to facilitate the tourism mining and industrial activities but also fishery activities. Therefore, these aspects form the basis for the regional development of Goa. The present study has been carried out only to focus on fishery resource and marketing development of Goa. The spatial planning strategy is required for the further development. Hence the geographical personality is essential to prepare a plan for future development.

The concept of spatial pattern is one of the most important concepts in geography. Search for pattern in the distribution of different phenomena on the surface of the earth has been one of the main pursuits of geographers, planners and diplomats/ policy makers. The distribution of fish landing centers and their size and nature are closely related to physical, environmental, economic factors and government policies. An attempt has been made to understand the spatial distribution of fish landing centers, i.e., marine as well as inland with several aspects, i.e., physical, geological, social and economic in the study area.

In the regional studies, the study of spatial as well as temporal variations in distributional pattern of fish landing centers, settlements and market is of great importance. In order to get a statistical measure of the pattern of distribution of fish landing centers, the nearest neighbor method of analysis has been used.

The study area has a favorable coastline and equable - equitable climate throughout the year. The growth of mangroves along the costs as well as brackish water bodies is influencing the growth of fish fauna on an unprecedented scale. As a result, the state has rich fish species diversity.

Inland fishery resources are an important aspect of study of fishery resources and their marketing in the study area. Inland fishery resources have been classified into three broad categories -

1)Rivers, Estuaries, Mangrove, Brackish Water and Khazan Lands,

2)Fresh Water Lakes, Reservoirs, Canals and Other impoundments,

3)Aquaculture.

Prawn farming, in Goa, is considered to be the most lucrative enterprise due to high market demand price and persistent global demand. Prawn farming development also holds immense employment potential to the local people. It provides direct employment to at lest 2 persons per one hectare cultured area. It also provides indirect employment to the extent of 1100 men per day for one hectare during construction, input supplies, handling of material, harvesting, etc.

Marine fishery resource of Goa comprising 104 kms of coastline characterized by innumerable creeks, bays, mangroves, swamps and coral reef, etc. It's a broken coastline, ideal for coastal navigation and development of fish landing centres, which is 1.25 percent of the country's total of 8192 kms. The continental shelf area extends upto 10,000 sq.kms 100 fathoms depth. Marine fishery resources (fisheries) have developed extensively over the years due to favourable conditions. According to Parulekar (1989), the potential pelagic fishery resources for the EEZ (Exclusive Economic Zone) are 69000 tons for the shelf, and 8000 tons for the oceanic area. The sustainable pelagic yield is projected as 46,560 tons per annum. Similarly, the potential demersal resources of EEZ are estimated to be 1,12,600 tons with a sustainable yield of 67,500 tonnes per year. Therefore, the total sustainable yield for both pelagic and demersal fisheries of Goa is projected to be 1,14,060 tonnes annually.

There are seven taluks located on the coast with 42 marine fishing villages engaged in extraction of marine fisheries. The highly productive fishable area in the sea is extended upto 20-40 fathoms and covers approximately a total area of the 2000 sq.miles. Major part of the fisherman population of 36894 is engaged in marine fishing activities to earn their livelihood.

In the year 1957, the Portuguese Government introduced mechanization with modern boats. Under the circumstances, the mechanization did not work properly, and the activity incurred heavy losses. The project was not successful, and it was only after the liberation of Goa in the year 1961 that the situation changed. The region was under the control of Central Government of India, as it was a Union Territory, i.e., Goa, Daman and Diu, and the Separate Department of Fisheries was created in the year 1963. Then the mechanization was given tremendous encouragement, thereby mechanized fishing came to be practiced on an intensive scale.

So far as fishery resource production and output values are concerned, normally the output values depend upon the fisheries production. Higher productivity, comparative records, a particular taluk is a leading producer of fisheries and output values in the state.

The main objectives of the present study is to assess the fishery resource development for the period from 2001-02 to 2009-10 based on the various attributes that have already been discussed for specific point of time. The various attributes that have been selected for the index are:

1) Percentage share of marine fish catching centres in each taluk to the state total.
2) Percentage share of population involved in marine fishing activities in each taluk to the state total.
3) Percentage share of mechanized boats used in fishing activities in each taluk to the state total.
4) Percentage share of traditional boats operated for fishing in each taluk to the state total.
5) Percentage share of production of marine fisheries in each taluk to the state total.
6) Percentage share of value of fisheries caught in each taluk to the state total.

After the Second World War marketing of fish has taken a new role in business activity. The fisheries have not become highly industrialized in all fishing nations. The new fishing techniques have been adopted to sell more fish. The modern fish marketing system lays emphasis on meeting the existing demand of fish, besides tapping the potential demand in the important markets. The marketing of any produce mainly depends upon the availability, consumption and demand. In Goa, traditional system of fish marketing is adopted. Modern marketing system as well as the fish marketing is normally done at the collection centres, which are mainly situated in the area of fish landing.

The following are the major constituents which influence fish trade greatly in the study area.

1) Role of Fishermen and Marketing Intermediaries

2) Price and Price Determination of Fish

3) Packaging

The early records of the human history show very clearly that salesmanship existed earlier in primitive form. Since barter system suffered from certain difficulties such as lack of double coincidence of wants, etc., buyers and sellers were in need of an article which would be double and acceptable to the community as a standard of value. In course of time the difficulty of buyers and sellers were solved when coins and paper currency came to be accepted as a medium of exchange.

Marketing of fish and fish products at the international level started in the year 1975. The frozen prawn export packed up, but under stiff competition and changing international tariffs, there was a setback. After 1986, export of marine products is a significant contributing factor in the state economy. Before 1988, export was quite negligible, the real export of fish and fish products commenced from the year 1990-91. The state is earning valuable income by exporting a variety of fish products and raw fish to various countries of the world, i.e., Malaysia, China, South Korea, Singapore, Mauritius, Japan, UK and Spain.

The marketing strategy is instrumental in the planning process to determine the effective measures in order to overcome the challenges identified in the fish marketing system. Maintaining high quality food and fish should be propagated as a strategy to stay ahead of other competing countries in the world market.

The traditional fishermen of Goa have been struggling for more that 15 years to prevent the ecological disaster of the Goan inshore waters, caused by trawlers and purseiners. Earlier fertilizer factory along the Salcete coast began dumping poisonous effluents into the sea. This pollution killed shoals of fish, the fishermen, along with the concerned villagers, students, teachers and other citizens, staged various protests against the factory demanding that it be closed, until it set up an effluent treatment plant. Then the voice of the people triumphed. This success gave a boost to the fishermen, and it became evident to them that, if they are united, they could solve the problem of trawlers and purseiners encroaching on their traditional fishing waters. With the active support of Mr. Mathany Saldanha, they came together to form the "Goenchea Ramponkarancho Ekvott" (The All Goa Fishermen's Union). They demanded the immediate ban of trawling and purseining in inshore waters and warned that if this demand was ignored, the Goans would export the produce, thus depriving the local population of one more source of fish.

The fisherwoman is a very familiar sight in Goa. The practice in most Goan fisher-folk families the husband to go out fishing and for the wife to take the responsibility of marketing the fishes.

The various problems of fishing industry in Goa such as, Environmental Problem, Socio Economic Problems, and Personal Problems, as well as markets. Necessary measures have been adopted by the Government of Goa and Fishery Co-operative Societies and Fish Traders Community to overcome the various problems in the study area.

6.2 FINDINGS

The following are the findings of research work carried out by the researcher.

1. The geomorphological and geological characteristics of Goa coast are ideal for fishing activities on account of numerous geeks, bays, landing centres and the growth of mangroves along the brackish water bodies.
2. Inland fish landing centres have been distributed on the geographical space in a clustered form. Hence, these centres provide fish catch to local areas as well as main markets.
3. Marine fish landing centres are distributed in the form of linear pattern in nature.
4. The fishermen are still using traditional boats for their fishing operation with risk.
5. Fishermen supply nearly 80 percent of their fish catch to market centres and the remaining 20 percent supplied to hotels, bar and restaurants, directly.
6. it has been observed that 55 percent of total fishermen practice there fishing activities in shallow waters, 27 percent in both shallow/deep sea and 18 percent of fishermen venture into the deep sea for fishing operations.
7. Fishermen community has an average literacy of 71 percent as they are becoming literate in the study area.
8. Fishermen community extract fish thoroughly their various nets. After sorting out fish, they throw a number of fish in to the sea which then float to the sea shore. Such kinds of fish cannot be marketed due to the spoilage and their tiny size.
9. The study area has plenty of fresh water bodies. It has been observed that inland fisheries have not been properly developed and fully exploited as compared to marine fisheries.
10. It has been observed that the prices of fish commodities are fluctuating and not reasonable due to seasonal variations in the study area.
11. The production of marine fisheries has been in consistent with that of the last

ten years, resulting into the shortage of fish commodity in the market centers.

12. During fishing ban period, i.e., 15th June to 31st July in the state, the neighboring regions such as Kerala, Tamil Nadu, Karnataka, supply on an average 1200 metric tons or 100 trucks loads of fish to goan consumers every day.
13. The fishing ban has been strictly observed during initial stage of S.W. Monsoon season for 45 days, in order to encourage spawning activities of fisheries, in view of Supreme Court guidelines.
14. The study has observed the fact that the fishermen community are facing a problem of labour supply in peak times.
15. The Canacona and Pernem talukas are underdeveloped with regard to basic infrastructure, i.e., scarcity of cold storage facilities, ramps and auction sheds affecting fisheries resource development.
16. Tiswadi taluka recorded the lowest fishery output that is 0.50 percent of total in the study area despite good infrastructure in the region during 2009-10.
17. Fish market centres are largely confined to census towns and cities rather than rural villages. And they have been distributed in the form of perfectly random spatial pattern in the study region.
18. There is no direct involvement of the Directorate of Fisheries, Government of Goa, towards marketing of fish products, and there is no direct control over the prices prevailing in the market throughout the year. The prices are generally decided by the fish traders and brokers, depending upon the demand and supply situation.
19. There are only 9 manufacturer exporters units involved in processing of marine fish and fish products. The concept of value added fishery products is not yet largely developed in the study area.
20. Fish trader and consumer ratio is quite higher in Bardez, Tiswadi, Salcete and Marmugao taluka, as they are grown as major fish market centres.

6.3 SUGGESTIONS

1. The breeding grounds of fish must be utilized properly to harness fishery resources, at optimum level and not destroy the nurseries of fish fauna.
2. Patrolling boats and monitoring fleet are necessary to curb illegal fishing during ban period, and they must be in good condition.
3. Sustainable fisheries demand awareness and self discipline. There should be allocation of sufficient funds for infrastructure and cold storage facilities in underdeveloped taluks.
4. Deep sea water pollution be controlled by government agencies.
5. Awareness needs to be created among fishermen community by organizing seminars, conferences and training programmes related to fishery resource development in a regional language.
6. Fish landing centers in Canacona taluk need to be properly utilized for increasing fish catch and production.

7. Enact and implement recovery plans for depleted fish species.
8. Develop social and economic incentives for sustainable development of fisheries.
9. Strictly execute the conservation and management of fishery resources - inland and marine fisheries under the provisions of Indian Fisheries Act.
10. Pernem and Canacona taluk needs to be given proper attention by Government of Goa to develop necessary infrastructure to promote fishery activities on large and commercial scale without affecting the existing environment.
11. The youth from KHARVI fisher fox/community should be encouraged by providing attractive incentives with measures of life risk to take up and continue fishing activities on commercial scale to ensure sustainable growth of fishery resources in the state.
12. Quality of marketing infrastructure needs to be developed in various taluks.
13. Fish traders should be encouraged by the Government through various schemes to promote wholesale and retail activities.
14. There should be effective monitoring on quality control of fish and fish products at fish processing units for export.
15. Prices of fish commodities in the daily market centers should be regulated and notified by the competitive authorities, i.e., Directorate of Fisheries, Government of Goa on regular basis. This will resolve the major problem of high soaring prices of various fish commodities which are out of reach of a common man in the study area.
16. Major jetties as well as others have excess trawlers. Therefore, the Directorate of Fisheries has to take a decision not to issue any more new licenses, except when trawlers have to be replaced due to sinkage and irreparable damages.
17. Inland fisheries have great potentiality to be developed and exploited in different parts of the study area. As the state holds a good number of inland water bodies and reservoirs.
18. The manufacturers of value added fishery products should be given promotional incentives to develop domestic as well as international marketing by the competitive authorities i.e. Marine Products Exporters Development Association (M.P.E.D.A.) and the Directorate of Fisheries, Government of Goa.
19. The drastic decline in fishery production and output values in the Tiswadi taluka during 2009-10, need to be reviewed by the Directorate of Fisheries, Government of Goa.
20. Fish trader and consumer ratio need to be improvised in moderately and sparsely populated regions through mobile retailing in the study area.

Thus, it is hoped that the present study of fishery resource development and marketing with a geographical perspective will reflect not only on the fishermen community, trader for their prosperity but also to contribute to their state economic development in an effective manner. The present study may be a model for the coastal areas.

BIBLIOGRAPHY

A.G. Untawale (2004) : Know our Shore Goa WWF for Nature India Goa State, Office Panaji.

Aarthi Shridhar (2003) : Research Paper "Conflict and related issues in Marine Resources Conservation".

Agarwal S.C. (1989) : Fishery Management. New Delhi: Ashish Publishing House.

Ahmed E.R. (1972) : Coastal Geomorphology of India. Orient, Longman, 222 pp.

Alexander Barbosa (2002) : Fishing for High Living. Goa Today Magazine.

Alwares Claude (2002) Fish Curry and Rice. The Goa Foundation Publication, Mapusa.

Ananth P.N. (2000) : Marine Fisheries Extension. New Delhi: Discovery Publishing house.

AnjaniKumar (2003) : Research Paper "WTO, Food Safety Standards and Regulatory Barriers: Implication for Indian Fisheries Exports". Implication for Indian Fisheries Exports, National Centre for Agriculture Economics and Policy Research, New Delhi.

Anjanikumar (2003) : W.T.O. Food Safety Standards and Regulatory Barriers. National Centre for Agricultural Economics and Policy Research, New Delhi.

Annual Report (2003-04) : The Marine Products Export Development Authority Govt. of India.

Annual Report (2004-05) The Marine Products Export Development Authority Govt. of India.

Annual Report (2005-06) : ICAR Research Complex for Goa. Indian Council of Agricultural Research, Ela, Old Goa, India.

Annual Report (2007-08) : Fishery Survey of India. Department of Animal Husbandry, Dairying and Fisheries Government of India, Mumbai-87.

Annual Report (2007-08) : The Marine Products Export Development Authority Govt. of Goa.

Applebaum W. (1961) : Teaching Marketing Geography by the Case Study Method. Economic Geography, Vol. 37, pp. 48-60.

Aravind A. Mulimani (2006) : Marketing Geography a Spatio-Functional Perspective. Premier Publication, Dharwad.

Arun Kumar Jha and Bijayananda Naik (2009) : Research Paper "Microbial Biotechnology: An Effective Means for Disease Prevention in Shrimp Farming Ponds". Fishing Chimes, June Vol. 29, No. 3.

Arvind M. Dwivedi (2004) : Data Bank on Geography. Anmol Publication Pvt. Ltd., New Delhi.

Avinash V. Raikar (2003) : Research Paper "Sustainable Development of The Marine Resources In Goa".

B.S. Negi (1997) : Geography of Resources. Kedarnath Publication, Meerut.

Bal D.V.K.V. Rao (1984) : Marine Fisheries. Tata McGraw Hill Publishing Company Ltd., Bombay.

Balbir Singh Negi (1977) : Human Geography. Kedarnath Ramnath, Meerut.

Balkrishnan S. (2008) : Research Paper "Women Employment in Fisheries Industries". Vol. 46, No. 23-24.

Bari Mulay and T.M. Patil Gujrathi (1988) : Commercial Geography.

Berry B.J.L. (1975) : Geography of Market Centres and Retail Distribution. Englewood Cliffs, N.J., Prentice Hall, pp. 1,2,3,125.

Bevinda Collaco July (2005) : Fishing Ban in Rough Waters Goa Today Magazine.

Bhatta Ramachandra (1996) : Role of Co-operation in the Management and Development of Marine Fisheries of Coastal Karnataka. In Rajgopalan (eds.) Rediscovering Co-operation: Strategies for the Models of Tomorrow. Anand Institute of Rural Management.

Bhattacharya Hrishikesh (2002) : Commercial Exploitation of Fisheries. Oxford University Press, New Delhi.

Bromely R.J. (1971) : Marketing in Developing Countries - A Review Geography, Vol. 56, pp. 124-32.

C. Gnaneshwal and C. Sudhakar (1997) : Published by: Kanthirava Offset Printer, Bhimavaram, Vijayalaxmi, Court Road, Anantpur, A.P.

C.B.L. Srivastava (2002) : Fishery Science and Indian Fisheries. Published by Kitab Mahal, 22-A Sarojani Naidu Marg, Allahabad.

Cherunilam Francis (1993) : Fisheries Global Prospective and Indian Development. New Delhi: Himalaya Publishing House.

Chorlay and Hagget (1968) : Models in Geography. Methuen, London.

Clark P.J and Evans E.C. (1954) : Distance to Nearest Neighbor as a Measure of Spatial Relationships in Population. Ecology, 35, pp. 445-53.

Clark P.J. and Evansi E.C. (1954) : Distance to Nearest Neighbour as a Measure of Spatial Relationship. Population Ecology, 35, pp. 445-53.

Cole and K.ing (1968) : Quantitative Geography. Techniques and Theories in Geography, Wiley, London.

Cushing D.H. (1975) : Marine Ecology and Fisheries. Cambridge University Press, p. 278.

D.D. Nambundiri (2007) : Research Paper "Post Harvest Technology of Farmed Fish". Fishing Chimes, January, Vol. 26, No. 10.

D'cruz Sharoan and Raikar A.V. (2004) : Ramponkars. In Goa between Modernization. Government and Deep Blue Sea. Economics and Political Weekly, Vol. 34, No. 20, May 15.

Das and Ashutosh Mishra (2003) : Impact of WTO on Indian Fisheries. College of Fisheries, G.B. Pant University of Agriculture and Technology, Pant Nagar (V.S. Nagar), (U.A.).

De Souza Teotonio R. (ed) : (1990) : Goa Through the Ages. Vol. II, An Economic History, Concept Publishing Company, New Delhi.

Directorate of fisheries Govt. of India (1991) : Fisheries Development in Goa. Government Printing Press, Panaji, Goa .

Directorate of Planning and Statistics (1992) : Goa Gazetteer, Govt. Printing Press, Panaji, Goa.

Dixit K.R. (1976) : Geomorphic Features of the West Coast of India between Bombay and Goa. Geographical Review of India, 38/3, pp. 260-281.

Dixit R.S. (1981) : Market Centres and their Spatial Development in the Umland of Kanpur. Kitab Mahal, Allahabad.

Dixit R.S. (1981) : Spatial Organization of Market Centres in Hamirpur District. Pointer Publisher, Jaipur.

Dwivedi S.N. and M. Devaraj (1983) : Tuna Fisheries of the India. Ocean and Development Prospects in Indian EEZ, Tuna Update: 6-11.

Ecoforum (2000) : Fish Curry Rice: A Citizens Report on the State of the Goan Environment. Mapusa : The Other India Press.

Economic Survey (2005-06) : Government of Goa, Directorate of Planning, Statistics and Evaluation, Panaji.

F.A.O. (2008) : Technical Guidelines for Responsible Fisheries Series, 4,5,11,12.

F.A.O. (2008) : Fisheries and Aquaculture Statistics 2006.

F.A.O. (2008) : Fisheries and Technical Papers, No. 510 & 513.

F.A.O. (2008) : The State of World Fisheries and Aquaculture.

F.A.O. (2009) : A Fisheries Manager Book Guide book 2nd Edition.

F.A.O. (2009) : Fisheries and Aquaculture Proceeding, No. 13.

G.P. Gandhi (2005) : Marine Product Export Post Tsunami. Facts for You.

George P.C., B.T. Antony Raja and K.C. George (1977) : Fishery Resources of the Indian Economy Zone. Souv. Integrated Fisheries Project Silver Jubilee Celebration:79-116, 219, GoG, Regional Plan for Goa (2001) A.D. Regional Plan Cell, Town and Country Planning, Government Printing Press, Goa, Panaji.

Goa Tourist Directory (1982) : Directorate of Tourism Government of Goa, Panaji.

Goa Tourist Directory (1982) : Directorate of Tourism, Government of Goa, Panaji.

Goh Cheng Leong and Gillian C. Morgan (1985) : Human and Economic Geography. Oxford University Press.

GoI (2001) : Agricultural Statistics at a Glance, 2001, Director of Economics and Statistics, Ministry of Agriculture, Government of India.

Gomantak Daily: 24 June 2010.

Government of Goa (1897) : The Indian Fisheries Act and The Goa Fisheries Rules. Govt. Printing Press, Panaji, Goa.

Government of Goa (2000) : Statistical Hand Book of Goa. Publication Division, Directorate of Planning, Statistics and Evaluation Panaji, Goa.

Government of Goa (2008) : Citizens' Charter for Captain of Ports and River Navigation Department, Govt. of Goa.

Government of Goa (2009) : Goa Economy in Figures. Directorate of Planning, Statistics and Evaluation, Panaji, Goa.

Govt. of Goa (2000) : The Indian Fisheries ACT 1987 and the Goa Fisheries Rules 1981. Govt. Printing Press.

Herald Daily: 7 January 2010.

Hodder B.W. (1965a) : Some Comments on the Origin of Traditional Markets South of Sahara. Transaction of Other Institute of British Geographers, 35, pp. 97-105.

Hodder B.W. (1965a) : The Distribution of Markets in Yoruba Land. Scottish Geographical Magazine, 18, pp. 57-91.

Hugar S.I. (1984) : Spatial Analysis of Market System in Dharwad District. Unpublished Ph.D. Thesis, Karnataka University, Dharwad.

Hugar S.I. (2000) : Traditional and Non-traditional Market Exchange - A Study in Spatial Development. G.K. Publishing House, Varanasi.

Satya Sundaram (1999) : January "Sea is the Limit". Facts for You, Market Survey.

Satya Sundaram (2001) : January "Marine Products - Exports hold the Key". Facts for You, Market Survey.

Satya Sundaram (2004) : Aquaculture: Fluctuating Fortunes. Facts for You.

Satya Sundaram (2006) : Freshwater Aquaculture No Longer a Backyard Activity. Facts for You.

Ibrahim P. (1992) : Fisheries Development in India. New Delhi: Classical Publishing Company.

Iyear S.D. and Wagle B.G. (1987) : Morphometric Analysis of the River Basin in Goa. Geographical Review of India, 49/2, pp. 11-18.

Joe D'Souza (2002) : Fishing Woes to the Fore. Goa Today.

Joseph K.M. (1985) : Marine Fishery Resources of India. In a System Framework of Marine Food Industry in India (Ed. E.R. Kulkarni and U.K. Srivastava), p. 90.

Joseph K.M., N. Radhakrishnan and K.P. Phili (1976) : Demersal Fishery Resources of South West Coast of India. Bull. Expl. Fish. Project.

K. Pau Biak Lun, and Ajit Kumar (2009) : Research Paper "Status of Aquaculture and Fisheries in Manipur State". Fishing Chimes, April, Vol. 29, No. 1.

K. Rama Krishnam Raju (1994) : Impact of Socio-Economic Development on the Fresh Water Lake Ecosystem - A Case Study of Kolleru Lake Region of A.P. India. Ph.D. Thesis, Andhra University, Visakhapatnam.

K. Vinod and B.K. Mahapatra (2007) : Research Paper "Umiam Reservoir Fisheries of Meghalaya". Fishing Chimes, January, Vol. 26, No. 10.

K.K. Khanna and Dr. V.K. Gupta (1994) : Economic and Commercial Geography. Sultan Chand and Sons Publishing.

K.N. Mohanta, S. Subramanian, N. Komarpant and A.V. Nirmale (2008) : Breeding of Gold Fish. Fishery Science Section ICAR, Ela, Old Goa. Sahyadri Offset System, Corlim, Goa.

K.P. Biswas (2009) : Fishes around Indian Ocean. Daya Publishing House, Devram Park, Trinagar, Delhi-35.

Kamat Nandkumar (2003) : Threat to Goa's Khazan Farms' in Navhind Times, Dated: December 15.

Kurian, John (1995) : Impact of Joint Venture on Fish Economy. Economics and Political Weekly, Feb. 11.

M. Das and Ashutosh Mishra (2003) : Research Paper "Impact of WTO on Indian Fisheries". College of Fisheries, G.B. Pant University of Agriculture and Technology, Pantnagar, U.S. Nagar (UA).

M.M. Prasad and J.K. Bandyopadhyay (2003) : Research Paper "Poverty Alleviation in Small-Scale Fishing Communities - A Review". Central Institute of Fisheries Technology, Burla Research Centre, Burla, Orissa.

M.S. Rao (1988) : Anmol's Dictionary of Geography. Anmol Publications, New Delhi.

M.V. Rama Prasad (2003) : February "Sea Food Needs Value Addition". Facts for You, Market Survey.

Majid Husain (1994) : Human Geography. Rawat Publications, Jaipur and New Delhi.

Majid Husain (1994) : Regional Geography. Anmol Publications Pvt. Ltd., New Delhi.

Mannash Choudhury (2009) : Research Paper "Marine Fisheries of West Bengal". Fishing Chimes, May, Vol. 29. No. 2.

Mineral and Export Intelligence (2009) : Goa Mineral Ore Exporters' Association, Goa.

Mohammad Shafi (2006) : Agriculture Geography. Dorling Kindersley (India) Pvt. Ltd.

Mrs. Usha Dessai (2006) : "Selection of Varieties of Prawns and Fish for Brackish Water Fish Farming".

Mrs. Usha Dessai and Mr. N.V. Verlekar : "Fish and Prawn Seed Resources of Goa". Directorate of Fisheries, Government of Goa.

Mulimani A.A. (1992) : The Geographical Analysis of Mineral Resource base Belgaum District for Development. M.Phil. Dissertation submitted to the Karnataka University, Dharwad.

N. Sarangi and Dr. J.K. Jena (2005) : Including Promising Species. The Hindu Survey of India Agriculture.

N.P.S. Varde (2003) : Research Paper "Management of Coastal and Marine Environment".

Naik S.K. (2003) : Research Paper "Challenges for Sustainable Capture Fisheries".

O. Henry Francis (2009) : Research Paper "Shrimp Culture under Contract Farming". Fishing Chimes, January and February, Vol. 28, No. 10/ 11.

P. Pravin and Saly N. Thomas (2009) : Research Paper "Studies on Drift Gillnet Fishing at Agatti Island (Lakshadweep). Fishery Technology, Vol. 46(1), pp. 7-14.

P.A. Koli (2008) : Research Paper "Problems of Fisheries Co-operatives".

P.C. Mahanta (2009) : Research Paper "Status Coldwater Fisheries Development in India". Fishing Chimes, April, Vol. 29, No. 1.

P.S.B.R. James (2009) : Research Paper "Responsible Fisheries, Key to Conservation, Management and Development of Fisheries in India". Fishing Chimes, July, Vol. 29, No. 4 .

Pandian J. (2001) : Sustainable India Fisheries. Published by National Academy of Agricultural Science.

Peter B. Mogle and Joseph J. (2000) : Upper Saddle River Fishes : An Introduction to Ichthyology. Prentice Hall Inc.

Phillippe Kotler (1994) : Marketing Management, Analysis Planning Implementation and Control. Prentice Hall India Ltd., New Delhi.

Prabhakar S. Angle (1983) : Goa - An Economic Review. The Goa Hindu Association, Kala Vibhag.

Prabhakar S. Angle (1994) : Goa Concepts Mis-concepts. The Goa Hindu Association, printed by Arun Naik, Akashar Pratiroop Pvt. Ltd., Wadala, Bombay.

Prakash Shinde (1992) : Commercial Geography. Sheth Publishers Pvt. Ltd., Educational Publishers "Pallavi-Kunj" Margao, Goa.

Prakash Shinde (2008) : Geography of Resources. Sheth Publishers Pvt. Ltd., Bombay.

Prithwish Roy (1992) : Economic Geography - A Study of Resources. Amitabha Sen, New Central Book Agency (P) Ltd., Kolkata.

R.B. Mandal (1988) : Systems of Rural Settlements in Developing Countries. Concept Publishing Company, New Delhi.

R.B. Singh (1986) : Geography of Rural Development Intra. India, New Delhi.

R.K Jana and Dr. R.K. Sena (2007) : Overwhelming Growth. The Hindu Survey of Indian Agriculture.

R.N. Tikkha (1981) : Physical Geography. Kedar Nath Ram Nath and Co., Delhi.

R.S. Biradar and Dr. S. Ayyappan (2006) : Enhancing Global Competition. The Hindu Survey of India Agriculture.

R.Y. Singh (1994) : Geography of Settlements. Rawat Publications, Jaipur.

Raikar A.V. (2003) : Research Paper "Sustainable Development of Marine Resources in Goa - What needs to be done?

Ramchndra Bhatta (2003) : Research paper "Status of Marine Resources Exploitation and Management in Karnataka".

Rashmi J. Desai (1999) : Environmental Studies. Vipual Prakashan, Mumbai.

Rashmi J. Desai (2007) : Geography of Resources. Vipual Prakashan, Mumbai.

Ravindra K. Majumdar (2006) : Value Added Fish Products. Fishing Chimes, Vol. 26, No. 4.

Rekha R. Gaonkar and R.B. Patil and Maria D.C. Rodrigues (2006) : Fishes and Fisheries. A.P.H. Publishing Corporation, 5 Ansari Road, Darya Ganj, New Delhi.

Resources Information Series (2008) : Fishery Survey of India. Opp. Microwave Station, Mormugao, Goa.

Role of Mineral Ore Exports in the Economic Development of Goa (2002) : National Council of Applied Economic Research, New Delhi.

Rubinoff J.A. (2002) : Pink Gold: Transformation of Backwater Aquaculture on Goa's Khazan Lands. Economic and Political Weekly, March, 31.

S. Subramanian and Neelam Komarpant (2003) : Research Paper "Application of Remote Sensing Techniques in Increasing Efficiency in Marine Resources".

S.A. Mohite and A.S. Mohite (2008) : Marketing of Fish and Fish Products - Dominant Role of Fishermen. Fishing Chimes, Vol. 28, No. 8.

S.B. Sawanth and Prof. A.S. Athavale (1994) : Population Geography. Sunil Anil Mehta Publishing House, Pune.

S.M. Hussain (2002) : Encyclopedia of Fish Culture. Rawat Publications, Jaipur.

Satya Sundarm (2001) : Exporters Hold the Key Facts for You.

Saxena H.M. (1975) : Geography of Transport and Market Centres : A Case Study of Hadaoti Plateau. New Delhi: S. Chand and Co.

Shamila Manteiro (2003) : Research Paper "Fisheries Resources of Goa : Its Management and Conservation".

Shamila Monteiro (2009) : Farming of Catla and Rohu in Brackish Water Ponds: A Success Story. Fishing Chimes, July, Vol. 29, No. 4.

Sherlekar S.A. (1986) : Marketing Management. Himalaya Publishing House, Bombay.

Shrivastava V.K. (1987) : Geography of Market Centres and Rural Development : A Study of the Rural Area of Terai Region. Inter India, New Delhi.

Silas E.G. (1969) : Exploratory Fishing by R.V. Varun, Bull. Cent. Mar. Fish. Res. Inst., pp. 12,86.

Silas E.G. (1986) : Cephalopod Resources Prospective, Priorities and Targets for 2000 AD. In: E.G. Silas (ed.) Cephalopod Bionomics, Fisheries and Resources of EEZ of Indian : Bull. Cent. Mar. Fish. Res. Inst., 37:174.

Silas E.G. and P.P. Pillai (1982) : Resources of Tunas and Related Species, Other Fisheries in the Indian Ocean. Bull. Cent. Mar. Fish. Res. Inst., 32:174.

Silas E.G. and P.P. Pillai (1985) : Indian Tuna Fishery Development Perspectives and Management Plan. In E.G. Silas (ed.), Tuna Fisheries of the Exclusive Economic Zone of India: Biology and Stock Assessment. Bull. Cent. Mar. Fish. Res. Inst., 36:193-198.

Sivprakasam T.E. and D. Sudarsan (1988) : Tuna Resources of the South West Coast of India as surveyed during 1986-87. Bull. Fish. Surv. India: 17, pp. 1-20

Sivprakasam T.E. and Patil (1986) : Results of Exploratory Tuna Longline Survey conducted in the Arabian Sea, South West Coast of India during 1985-86. Occ. Pap. Fish. Surv. India, No. 3, p. 10.

Smitha R. Nair and S.K. Pandey (2003) : Research Paper "Cooperative Interventions for Enhancing Fisheries Development. The Case with Indian Fisheries.

Sreenivasan P.V. and R. Sarvenan (1990) : On the Cephalopods collected during the Exploratory Survey by FORV Sagar Sampada in the Andaman-Nicobar Seas. Proc. First Workshop Scient. Resul. FORV Sagar Sampad, 5-7 June, 1989:409-413.

Srinivasa Rao P. and Umesha D. (2009) : Research Paper "Fish Vaccination for Sustainable Aquaculture Production". Fishing Chimes, July, Vol. 29, No. 4.

State of Goa (2008) : Indicators of Socio-Economic Development. Directorate of Planning, Statistics and Evaluation, Government of Goa, Panaji, Goa.

Statistics of Marine Product Exports (2005) : The Marine Products Export Development Authority. Ministry of Commerce and Industry, Govt. of India.

Suda A. (1974) : Recent Status of Resources of Tuna Exploited by Longline Fishery in Indian Ocean. Bull. Far. Fish. Res. Lab. 10:29-64.

Sudarsan D. (1978) : Result of Exploratory Survey around the Andaman Islands. Bull . Expl. Fish. Proj., 7:43.

Sudarsan D., M.E. John and Antony Joseph (1987) : An Assessment of Demersal Stocks in the South West Coast of India with Particular Reference to the Exploitable Resources in Outer Continental Shelf and Slope. National Symposium on Research and Development Marine Fisheries, Mandapam Camp. Bull. Cent. Mar. Fish. Res. Inst., 44(1): 266-274.

Sudarsan D., M.E. John and V.S. Somavanshi (1988) : Observation on Abundance, Distribution and Seasonality of Yellow fin tuna in India Seas as revealed by Exploratory Surveys during 1983-88. IPTP Coll., Vol. Work. Doc., 3:251-261.

Sudarsan D., M.E. John, A.K. Bhargava and S.M. Patil (1990) : Tuna Longline Fishery in India. IPTP Expert Consultation on Stock Assessment of Tuna in the Indian Ocean, Bangkok, 2-6, July, 1990.

Sudarsan D.T., M. Sivaprakasam, M.E. John and V.S. Somavanshi (1989) : Assessment of Oceanic Tuna and Allied Fish Resources in Indian EEZ. National Conference on Tunas, Cochin, 21-22 April, 1989.

Sudarsan D.T., M. Sivaprakasam, M.E. John, V.S. Somavanshi, K.N.V. Nair and Antony Joseph (1988) : An Appraisal of the Marine Fishery Resources of the Indian Exclusive Economic Zone. Bull. Fish. Surv. India, 18:81.

Sulochana Chari and M.S. Saxena (2009) : Research Paper "Fishing Ponds Fertilized with Different Organic Manures". Fishing Chimes, March, Vol. 28, No. 12.

Sulochanan P., M.E. John and K.N.V. Nair (1986) : Preliminary Observations on Tuna Resources of the Arabian Sea with particular reference to Distribution Pattern of Yellow fin tuna. Thunnus albacores (BonnAaterre), Bull. Fish. Surv. India, 14:21-32.

Surapa Raju (2008) : Fish Retailers. Fishing Chimes, Vol. 28, No. 7.

T.C. Sharma and O.C. Countinho (1983) : Economic and Commercial Geography of India. Vikas Publishing House Pvt. Ltd.

The Times of India, Goa 18 August 2010.

The Times of India, Goa 2 August 2010.

The Times of India, Goa 31, July 2010.

The Times of India, Goa 7 March 2011.

Tourist Statistics (2008) : Department of Tourism, Government of Goa, Patto, Panaji-Goa.

Truman A. Hartshorn and John W. Alexander (1994) : Economic Geography. Published by Prentice Hall of India Pvt. Ltd., M-97 Connaught Circus, New Delhi.

V. Vivekanandan (2003) : Research Paper "Excessive Mechanization of Fishing Boats in Tamil Nadu and its Consequence for its Artisanal Fishermen and Neighbouring States. South Indian Federation of Fishermen Societies.

Varghese P. Oommen (1980) : Result of the Exploratory Fishing in Quilon Bank and Gulf of Mannar. Bulletin No. 4, IFP.

Varghese P. Oommen (1985) : Deep Sea Resources of the South West Coast of India. Bulletin No. 11, IFP.

W.H.L. Allsopp (1985) : Fishery Development Experiences. Published by Fishing News Books Ltd., 1 Long Garden Walk, Farnaham Surrey, England.

Wagle B.G. (1982) . Geomorphology of the Goa Coast. Proceeding of the Indian Academy of Sciences (Earth and Planetary Sciences), 91, 105-117.

Wanmali S. (1981) : Periodic Markets and Rural Development in India. B.R. Publishing Corporation.

William F. Royce (1996) : Introduction to the Practice of Fishery Science. Academic Press, San Diego, New York.

Y.S. Yadava (2006) : "Exploiting Deep-Sea Avenue". The Hindu Survey of Indian Agriculture.

Z.A. Ansari and S.G. Dalai : "Over Exploitation of Fishery Resources, with Particular Reference to Goa". Directorate of Fisheries, Government of Goa.

Z.A. Ansari and S.G. Dalal and R.A. Sreepada (2003) : Research Paper "Ecosystem based Fishery Management for Sustainable Coastal Fisheries in Goa West Coast of India. National Institute of Oceanography, Dona Paula, Goa, India.

www.ingramcontent.com/pod-product-compliance
Ingram Content Group UK Ltd.
Pitfield, Milton Keynes, MK11 3LW, UK
UKHW021947270726
14060UKWH00002B/399

9 789390 371297